S0-DUV-768

Lobsters, Crabs, Shrimps, and Their Relatives

BOOKS BY RICHARD HEADSTROM

The Beetles of America
Spiders of the United States
Your Insect Pet
Frogs, Toads, and Salamanders as Pets
Whose Track Is It?
Lizards as Pets
A Complete Guide to Nests in the United States
Nature in Miniature
Adventures with Freshwater Animals
Adventures with Insects
Adventures with a Hand Lens
Garden Friends and Foes
Birds' Nests of the West
Birds' Nests
The Living Year
Adventures with a Microscope
The Story of Russia
The Origin of Man
Families of Flowering Plants

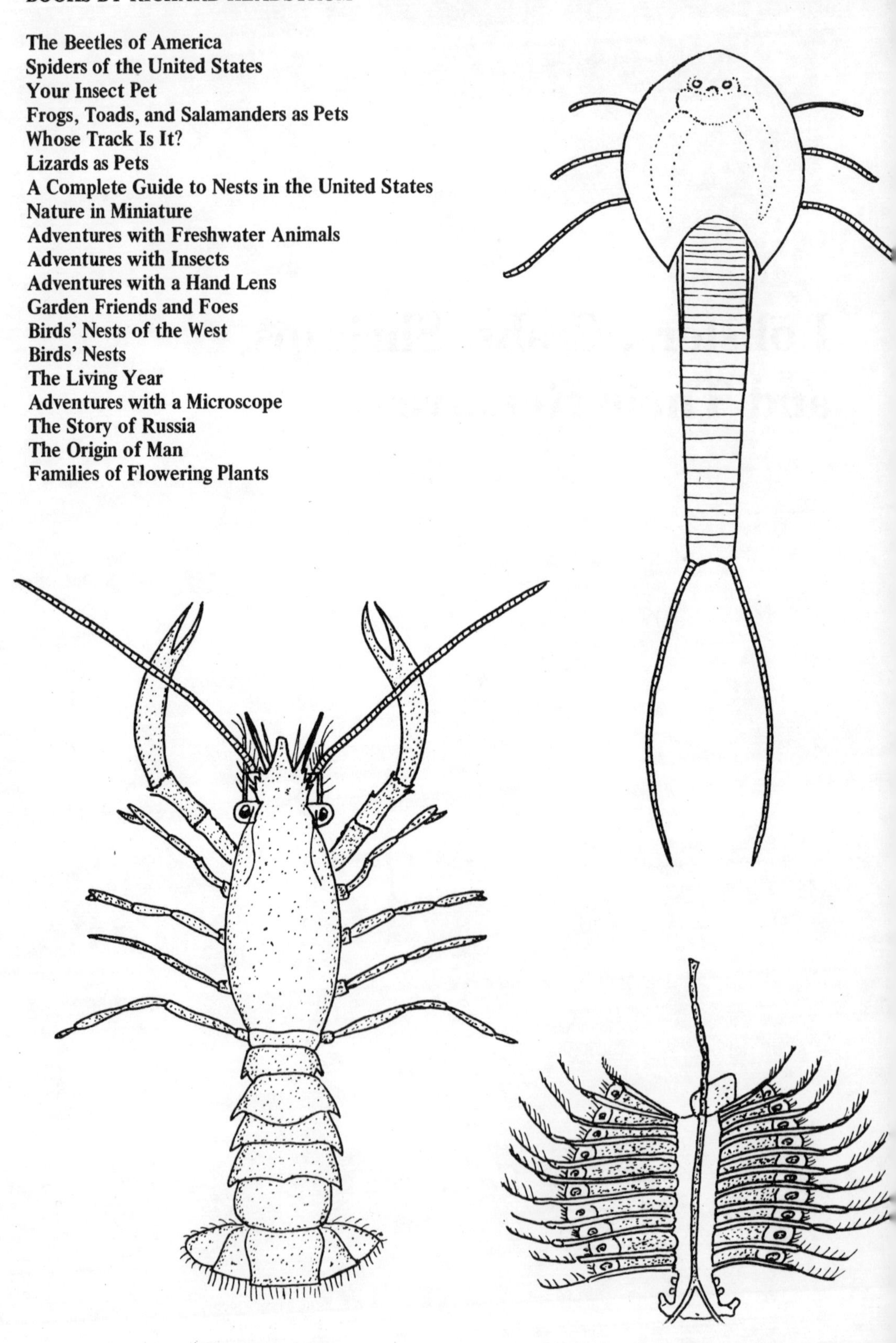

LOBSTERS, CRABS, SHRIMPS, AND THEIR RELATIVES

Richard Headstrom

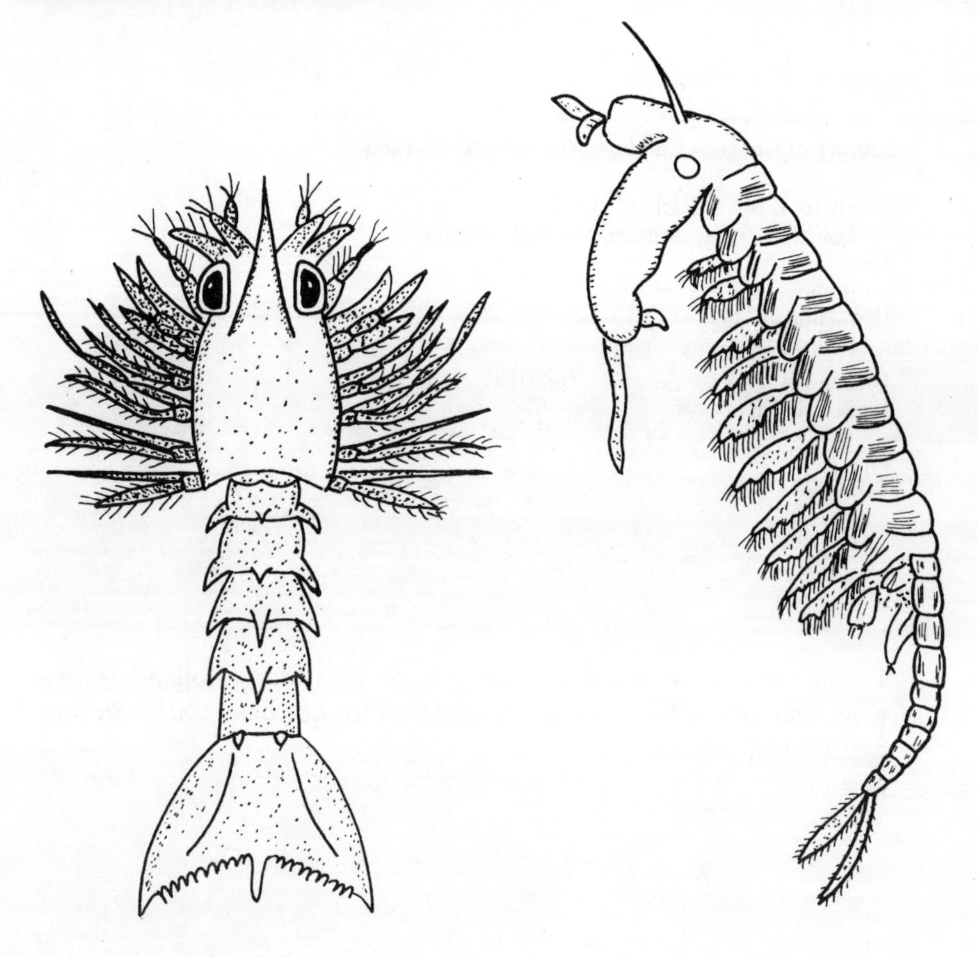

SOUTH BRUNSWICK AND NEW YORK: A. S. BARNES AND COMPANY
LONDON: THOMAS YOSELOFF LTD

A. S. Barnes and Co., Inc.
Cranbury, New Jersey 08512

Thomas Yoseloff Ltd
Magdalen House
136-148 Tooley Street
London SE1 2TT, England

Library of Congress Cataloging in Publication Data

Headstrom, Birger Richard, 1902-
Lobsters, crabs, shrimps, and their relatives.

Bibliography: p.
Includes index.
1. Crustacea. I. Title.
QL435.H4 595'.3 76-50193
ISBN 0-498-01764-8

I want to express my deep gratitude to Miss Mary Holland, who so graciously offered to provide photographs for *Lobsters, Crabs, Shrimps, and Their Relatives.*

Printed in the United States of America

To Ruth
in appreciation

Contents

Preface

The nearest that most of us get to a lobster is when one is served to us in a restaurant. And much the same can be said for the crab and shrimp. That these animals are good eating seems to limit our interest in them except for those engaged in catching them for the market and the few zoologists who have adopted them as their particular sphere of study and research. Which should not be the case, for, apart from their market value, they and their relatives are as interesting a group of animals as any others to be found in the animal kingdom. Moreover, they occupy a significant place in Nature's scheme of things.

Much of what we know about them is to be found in monographs, papers in scientific journals and the like, and in textbooks on zoology, most of which are not readily available to the general reader. True, they are briefly mentioned in many popular books on natural history, but even so they are given rather scant treatment in them. Hence a book dealing exclusively with lobsters, crabs, shrimps, and their relatives would seem to fill a definite need. Which is the reason for the present volume. And a valid reason, or so I believe, for imposing another book on a market already flooded with books and more appearing each year.

The present volume is popular in treatment and written for the general reader who, it is hoped, may find it an interesting and enjoyable reading experience.

Lobsters, Crabs, Shrimps, and Their Relatives

1
The Place of Lobsters, Crabs, Shrimps, and Their Relatives in the Animal Kingdom

Zoologists classify lobsters, crabs, shrimps, and their relatives as arthropods. An arthropod (which is from the Greek *arthron,* joint; *pous,* foot) is an invertebrate animal, that is, one without a backbone, having jointed appendages. Other arthropods or joint-footed animals are insects, spiders, mites, scorpions, centipedes, and millipedes, the last two being the hundred-legged and thousand-legged animals with which many of us are familiar, although neither has a hundred legs nor a thousand legs. Merely a glance at any of these animals will show that in respect to their appendages they are all structurally alike though they may differ in other ways.

An arthropod is said to be a bilaterally symmetrical animal, in other words it is so formed that its chief organs are generally arranged in pairs on either side of an axis passing from the head or anterior end to the tail or posterior end. Put differently, there is only one plane through which the body can be divided into similar parts. In an arthropod we may distinguish an upper or dorsal surface and a lower or ventral one as well as a right side and a left one. An arthropod, moreover, consists of a longitudinal series of segments, on all or on some of which is a pair of appendages, and has a hard outer covering that is flexible at intervals to provide movable joints. Such a hard outer covering is called an exoskeleton, since it protects and supports the internal organs.

In the modern system of classification, the animal kingdom, which consists of all known animals, is divided into a number of main divisions called phyla (sing. phylum), which in turn are further subdivided. These subdivisions in order are class, order, family, genus, and species. Hence if we were to classify an arthropod, such as the lobster for instance, we would proceed as follows:

Phylum Arthropoda
  Class Crustacea
    Subclass Malacostraca
      Order Decapoda
        Family Nephropidae
          Genus *Homarus*
            Species *americanus*

The scientific name of the American lobster is *Homarus americanus.*

It will be noted from the above that the lobster belongs to the class Crustacea. So, too, the crabs, shrimps, and all their relatives such as the barnacles, crayfishes, fairy shrimps, sow bugs, and beach fleas. Most members of the class Crustacea are small animals, under half an inch in length. The lobsters and the crabs are the giants.

The General Characteristics of the Crustaceans

The word *crustacean* comes from the Latin *crusta* or shell and was originally used to designate an animal having a hard but flexible covering or shell as contrasted with an animal, such as the oyster or clam, having a hard but brittle covering or shell. But since nearly all arthropods have a hard, flexible covering it has become necessary to use more distinctive criteria for designating an animal as a crustacean. As a matter of fact the crustaceans show such a diversity of structure and habits that it is well nigh impossible to come up with a brief definition that will fit all of them. But generally speaking we might distinguish them from other arthropods by saying that most of them are aquatic, breathing by gills or by the general body surface, and that they have two pairs of preoral antennae (that is, antennae situated in front of, or anterior to, the mouth) and at least three pairs of postoral appendages that act as jaws. The preoral appendages may be locomotor or prehensile in function or may be absent altogether, and some or all of the mouth parts may have been arrested in their development or failed to have developed at all. There are some crustaceans like the terrestrial sow bugs and the land crabs that have special organs for breathing atmospheric oxygen, and then there are some extremely modified parasitic forms that as adults show little evidence of being crustaceans or even arthropods though their true character is revealed in their larval stages.

Thales, the Greek philosopher, once said that he believed all life to have originated in the water. We might sort of paraphrase this statement by saying that the earliest crustaceans lived in water, more specifically in salt water, since the seas were more conducive to life than the land and necessitated the least number of adjustments on the part of animal forms to live in them. For the seas provide a relatively constant salt and oxygen content and a more even temperature throughout the year and, furthermore, salt water is buoyant and offers greater support. Although some crustaceans live in fresh water they are predominantly marine and are so abundant in the oceans that they have been called "the insects of the sea."

That the crustaceans, which are essentially marine animals, have done as well in fresh water as they have may seem somewhat surprising, for fresh water is a rather difficult medium in which to live. And yet it is perhaps not so surprising, for animals are capable of great adjustments. The transition from a marine environment to a freshwater one had to be by means of the rivers that were directly connected with the oceans, but to be able to invade the rivers they first had to adjust themselves to the lower salt content (it being remembered that the salt concentration of animal tissues is nearer that of sea water than fresh water) and then secondly to find a way to withstand the river currents, which had a tendency to sweep them out to sea, a factor that, incidentally, has prevented other invertebrate animals from becoming established in fresh water. The adults apparently were successful in combatting the downstream currents, otherwise they could not have succeeded in ascending the rivers and remaining in them, but their small and fragile larvae were unable to do so. The crustaceans that were eventually successful in remaining in a freshwater habitat were those that suppressed their free-swimming larval stages and instead evolved young that were miniature facsimiles of themselves and that were capable of maintaining themselves against the river currents.

In addition to the problems of adjusting to the lower salt content and the river currents, the crustaceans had still another one: that of adjusting to the violent temperature fluctuations to which the ponds and rivers are subject. They met this problem by developing thick-shelled eggs that resist drying and freezing.

To adapt themselves to a land habitat posed an even greater problem, for not only are temperature fluctuations on land more extreme but there is ever present the threat of drying up. Then, too, they had to make a change in their respiratory structures so that they could breathe atmospheric oxygen. Some of them, such as the land crabs and the "wood lice," succeeded in doing so but only, however, because they live in moist places.

The External Anatomy of Crustaceans

Body Form Most crustaceans have an elongated body composed of a longitudinal series of segments or somites. In many instances, many of the segments have become fused together so that the result has been a shortening of the body (Fig. 1).

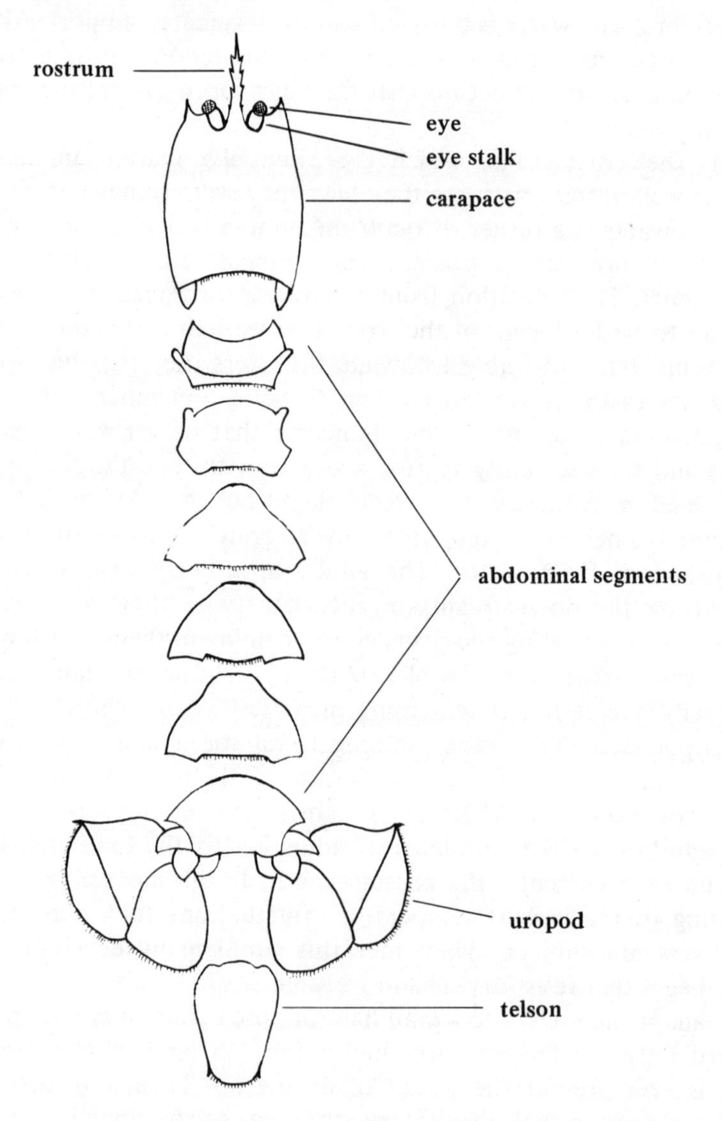

Figure 1. External Dorsal Anatomy of a Lobster

Integument All crustaceans have an outer covering or integument that is composed of a shell-like cuticula, which forms the entire outer surface, and beneath it a layer of glandular cells, called the hypodermis, which secretes it. The cuticula in turn is made up of two layers, in some places three; an outer thin, waxy, waterproof layer, an inner layer of several substances, the best known being chitin, a nitrogeneous polyssacharide, and a middle layer between these two of mostly lime salts (Fig. 2). Chitin is a very resistant substance and is insoluble in water, alcohol, dilute acids, alkalies, and the digestive juices of many animals. At one time it was believed that the hardness of the cuticula was due to chitin, but it is now known to be due to the lime salts in the middle layer, the chitin instead providing flexibility or elasticity. In short the cuticula occurs as hard or rigid areas and as flexible membranes or joints between them. Thus the integument of the crustaceans provides the animals with a protective armor without sacrificing mobility, an armor that is so much better than that of such animals as the snails and clams, which must carry around a heavy, cumbersome shell that limits their movements.

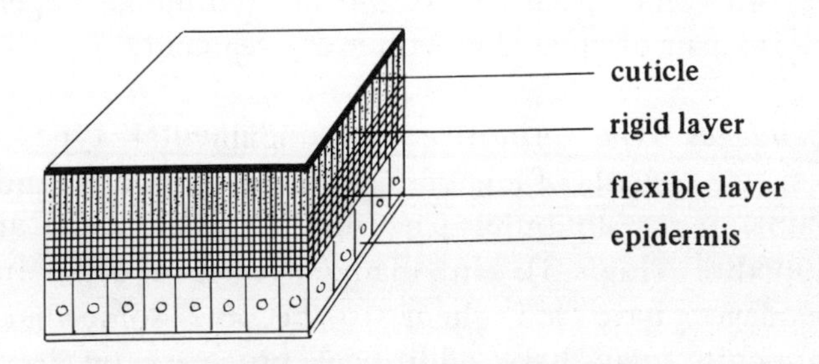

Figure 2. Body Wall of an Arthropod

Exoskeleton Since the integument provides a supporting framework for the internal organs and a surface for the attachment of muscles, it is appropriately called an exoskeleton, in other words an outer skeleton or a skeleton on the outside. In contrast is the internal skeleton of the vertebrates, which is called an endoskeleton. Although it was once believed to be responsible for the hardness of the cuticle, which is now known not to be the case, the word *chitinous* is used to distinguish the exoskeleton of the crustaceans as well as other arthropods from the external supports of other animals. This chitinous exoskeleton consists of a series of segments (somites) that may be either movably articulated or more or less fused together and all of which may bear appendages except the last one, the telson, which actually is not a true segment (Fig. 1). Sometimes the telson has a pair of prongs called the caudal furca.

Body Regions In the crustaceans three body regions are recognized: head, thorax, and abdomen. In some species, as the lobster, the head and thorax are fused, forming what is called the cephalothorax; in others, such as the crabs, the abdomen may be much reduced. In the crustaceans the part of the integument covering the cephalothorax entirely or only partially is known as the carapace (Fig. 1). In some species it may loosely envelop the limbs and more or less form a sort of bivalve shell; in others it protects the gills on either side.

Head The head consists of at least five segments that are fused and typically bears a pair of jointed appendages that may be either sensory or function in feeding. The head also bears a pair of compound eyes.

Thorax In the crustaceans the number of segments in the thorax varies greatly. The number of appendages also varies, from two the smallest number to sixty the largest.

Abdomen As in the thorax, the number of abdominal segments also varies. In some species there are no abdominal appendages; in others there is a pair of appendages on every segment.

Appendages The primitive or fundamental type of a crustacean appendage consists of a basal part, the protopodite, and two segmented, terminal parts, an outer part, the exopodite, and an inner one, the endopodite (Fig. 3). In many species special functions of the appendages have led them to become somewhat modified. The protopodite may have additional processes on its outer and inner margins, known as exites and endites. Some of the exites often function as gills and the endites of the appendages near the mouth often form jaw-processes.

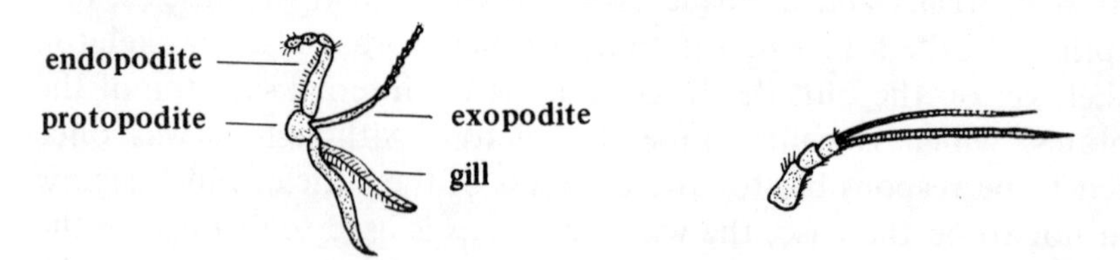

Figure 3. 2nd Maxilliped

Figure 4. 1st Antenna or Antennule

There are five pairs of appendages present on the head: antennules, antennae, mandibles, and first and second maxillae. The antennules (Fig. 4) differ from the other appendages in that they are one-branched (in the earliest larvae and in many adults) or uniramous, whereas the other appendages are typically two-branched or biramous, but in some instances they may be three-branched or triamous. The antennules are essentially sensory in function but in some species they may serve as organs of attachment.

In the adult crustaceans, the antennae (Fig. 5) are chiefly sensory but in the parasitic forms that may serve as organs of attachment. In the nauplius larva they are used in swimming and sometimes in chewing. Like the antennae the mandibles also serve as swimming organs in the nauplius larva, and they also perform this same function in some adults, but in most adult crustaceans they have become modified to act as chewing organs. In these the exopodite has become lost, the endopodite forms a palp or has become lost altogether, and the protopodite has become enlarged to form the body of the mandible (Fig. 6). In the parasitic species with sucking mouthparts, the mandibles have become modified to form piercing lancets that are enclosed in a sheath formed by the upper and lower lips.

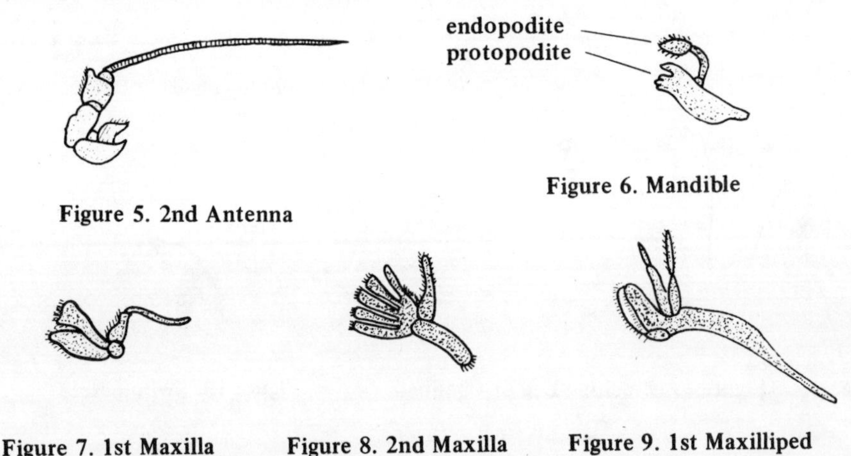

Figure 5. 2nd Antenna

Figure 6. Mandible

Figure 7. 1st Maxilla

Figure 8. 2nd Maxilla

Figure 9. 1st Maxilliped

The first and second maxillae, which are used to convey food to the mouth, are nearly always leaflike with lobes on the protopodite, the exopodite generally being present and the endopodite being reduced to a palp or absent (Figs. 7, 8).

In some species of crustaceans, the appendages of the thorax are nearly all similar and leaflike but in others they have become modified to function in food handling, the maxillipedes (Figs. 3, 9, 10), or as walking legs (Fig. 11), some of which end in pincers (Fig. 12). In most crustaceans the appendages of the abdomen, known as swimmerets (Fig. 13), are two-branched or biramous. They function as swimming organs in the more primitive forms as well as in some adults, and in others they are used for respiration. In some species they are absent. The appendages are primarily locomotory and sensory organs, but they also perform other functions.

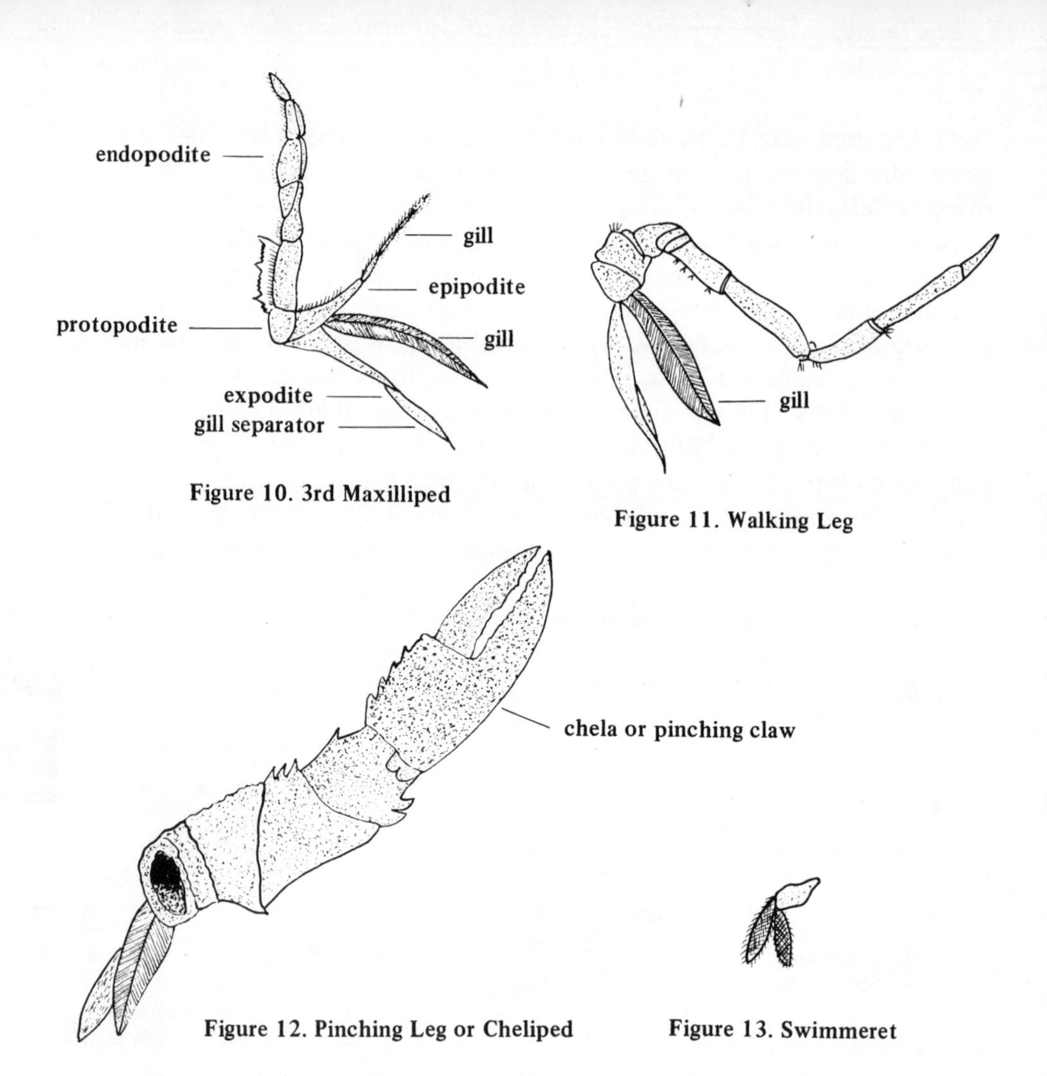

Figure 10. 3rd Maxilliped

Figure 11. Walking Leg

Figure 12. Pinching Leg or Cheliped

Figure 13. Swimmeret

The Internal Anatomy of Crustaceans

Digestive System The digestive system of the crustaceans consists essentially of a long tube that curves or bends down in the anterior part of the body to the ventrally situated mouth and that in a few instances is twisted or coiled (Fig. 14). We may recognize three sections: the foregut, midgut, and the hindgut, with an anterior opening, the mouth, which is located between the jaws, and a posterior opening, the anus, which in some species is located on the telson. This system may be somewhat modified in the various groups. In the more primitive forms there are spines and hairs on the lining of the foregut that help to grind the food; in others a series of plates operated by muscles performs the same service. Such grinding structures attain their greatest development in the decapods, where three hard teeth, collectively known as the "gastric mill," move one upon another and are most effective in triturating the food. Digestive glands, which are paired tubular outgrowths from the midgut, secrete juices that convert the solid food into soluble substances and thus promote absorption. In some instances the tubular outgrowths are more or less branched and form a massive "liver" or what may be called a hepatopancreas. Undigested food materials are gathered together and expelled through the anus as feces (Fig. 14).

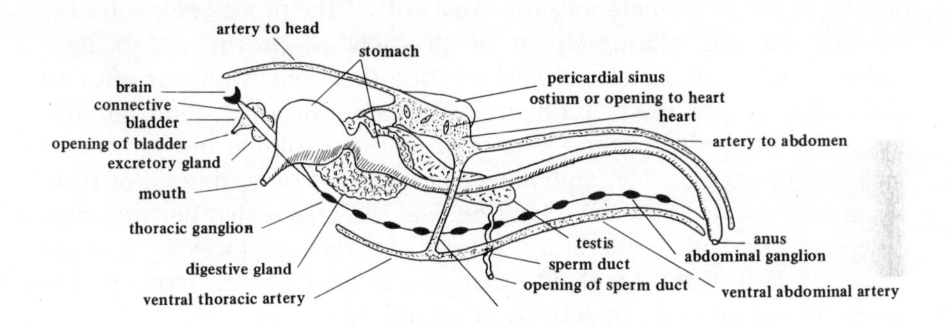

Figure 14. Internal Anatomy of a Lobster

Circulatory System The circulatory system of the crustaceans is composed essentially of a heart (though some species lack a heart), several blood vessels (which may be absent), and a blood plasma. The heart lies in a sinus or cavity, called the pericardial sinus, with which it communicates by means of openings or ostia. In some primitive forms the heart is long and tubular, with a pair of ostia in each segment, but it becomes progressively shorter through the various groups of crustaceans until it eventually becomes quite short with only two or three ostia (Fig. 14).

The blood plasma into which the absorbed food passes is an almost colorless liquid, and its principal function is that of transportation, carrying food materials to various parts of the body, oxygen from the gills to the various tissues, carbon dioxide to the gills, and waste materials of metabolism to the excretory glands. By rhythmic pulsations of the heart, the blood is forced through the blood vessels, or arteries, into spaces called sinuses or "blood cavities" that lie throughout the body or tissues and in the absence of arteries directly into the sinuses. In the sinuses the blood bathes the various organs and is then returned to the pericardial sinus, where it enters the heart to circulate again. Valves in the arteries and in the ostia prevent the blood from flowing back. In such crustaceans that lack a heart the blood is driven throughout the body by movements of the body, the alimentary canal or food tube, and the appendages such as the limbs. Since the blood is not confined entirely in blood vessels this type of circulatory system is known as an open circulatory system in contrast to the closed circulatory system such as is found in the vertebrates in which the blood is confined at all times in the arteries, capillaries, and veins.

Excretory System Metabolism, mentioned in the preceding section, can be defined as the sum total of all the reactions that take place within the protoplasm of an organism and that are mainly chemical in nature or, put differently, metabolism consists of all the processes involved in building up and tearing down of protoplasm, protoplasm being a complex physiochemical colloidal solution that constitutes the physical basis of life. The waste products of metabolism are carbon dioxide, certain soluble nitrogenous salts such as urea, soluble inorganic salts such as sodium chloride, and water. The process of getting rid of these waste materials is known as excretion, and the organs that perform such a service are called excretory organs. In the crustaceans they are generally two pairs of glands that are located at the bases of the antennae and maxilla respectively (Figs. 14, 15).

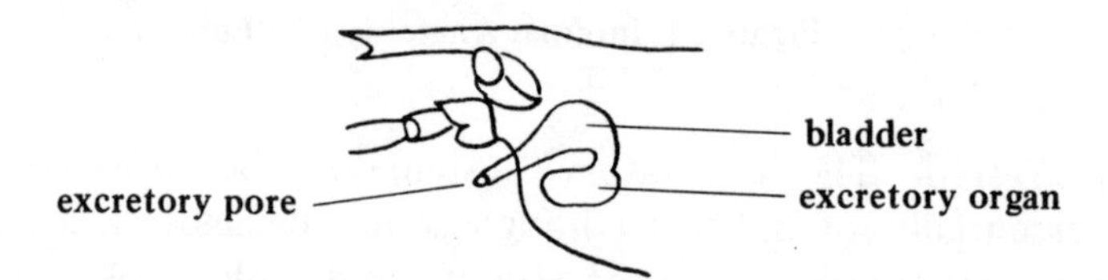

Figure 15. Excretory Organ of the Lobster

Respiratory System Respiration, which may be defined simply as the actual use of oxygen by a living cell, actually involves the processes by which a living cell obtains oxygen from the air or water, the transport of such oxygen to the cell, the use of the oxygen by the cell, and the discharge of carbon dioxide that results from the chemical reactions that take place within the cell. In the mammals the organs that effect an exchange of oxygen and carbon dioxide with the air are known as lungs; in the crustaceans they are known as gills, which are thin-walled extensions or projections from the legs or the sides of the body (Fig. 11).

Muscular System Movements in animals are due principally to muscles, a muscle being defined as a bundle of special tissue that can be extended or contracted (Fig. 16). In the lobster, for instance, the largest muscles are located in the abdomen, where they are used to bend the abdomen to produce locomotion. Muscles of considerable size are also located in the thorax and within the tubular appendages, especially the chelipeds or claws.

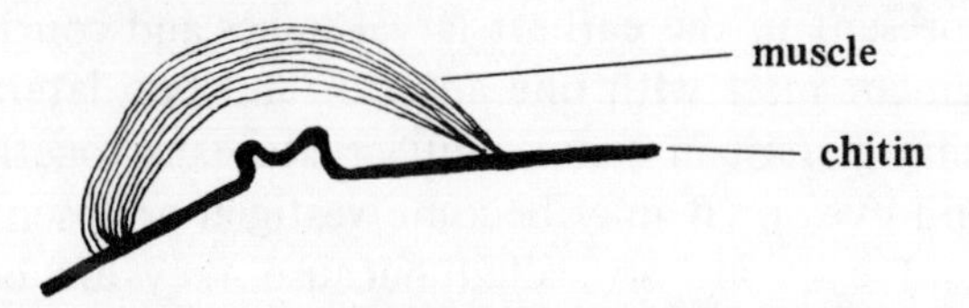

Figure 16. Part of an Arthropod Limb showing Chitin and Muscle

Nervous System The nervous system in the crustaceans consists of a nerve cord that lies ventrally along the length of the body and associated with it a series of ganglia (sing. ganglion) that are simply masses of nerve cells, one to each segment, and from which nerves extend to parts of the body. One such ganglion is located in the dorsal part of the head and is generally known as the "brain." It is connected to the nerve cord by two connectives that pass around the food canal and from it nerves extend to the eyes and antennae (Fig. 14).

Sense Organs The sense organs are organs that are connected to the nerve cord by means of nerves and that are capable of responding to

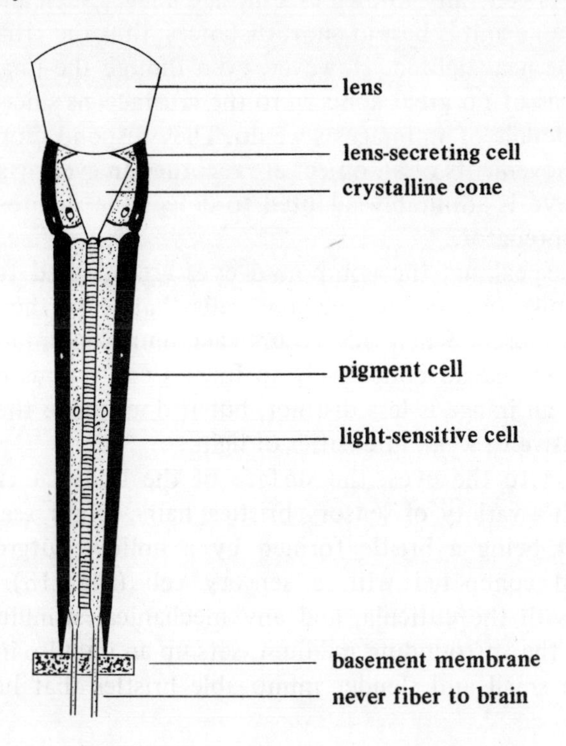

Figure 17. An Ommatidium of a Compound Eye

external stimuli and transmitting impulses to various parts of the body. The most important sense organs are the eyes. When present in crustaceans the eyes may be of two kinds: an unpaired median eye and a pair of compound eyes. The unpaired median eye, called a nauplius eye, is usually present in the earliest larval stages and consists typically of a central pigment mass with one anterior and two lateral groups of visual cells. It may persist in some adult crustaceans together with the paired compound eyes or it may become vestigial or absent altogether.

The compound eyes are so called because they are composed of hundreds or thousands of visual units that are known as ommatidia. They lie side by side but are separated from one another by dark pigment cells. An ommatidium (Fig. 17) is simply a bundle of cells consisting of refractive bodies that transmit the light rays and condense them upon the retina, which is composed of a layer of light-sensitive cells that are continued at their lower ends into nerve fibers that enter the central nervous system.

The external convex surface of a compound eye is covered by a modified section of the transparent cuticle called the cornea. The cornea is divided by a large number of fine lines into four-sided areas termed facets, each facet being but the external part of an ommatidium. Each ommatidium responds simply to a fragment of the total field, and these fragmentary images are fitted together into a single general picture that is generally known as a mosaic image. Such an image is not a very good one and is best at short distances; thus the crustaceans may be said to be near-sighted. However, even though the image may be a poor one, it is of no great concern to the crustaceans since they do not respond to details of an image as we do. They respond more to motion, and as the movements of an object are recorded in every visual unit, the compound eye is admirably adapted to detect the slightest movement of prey or a predator.

Generally speaking, the compound eyes are adapted to see in dim light by a migration of the pigment cells that leave the sides of the ommatidia exposed. When this occurs each ommatidium no longer acts separately but instead collectively to form a continuous image on the retina. Such an image is less distinct, but it does make the crustaceans more responsive to weak intensities of light.

In addition to the eyes, the surface of the body of crustaceans is covered with a variety of sensory bristles, hairs, spines, scales, and pits, the simplest being a bristle formed by a hollow outgrowth of the cuticula and connected with a sensory cell (Fig. 18). The bristle articulates with the cuticula, and any mechanical stimulus, such as a vibration in the surrounding medium, sets up an impulse in the sensory cell. Certain small and slender immovable bristles that have thin and

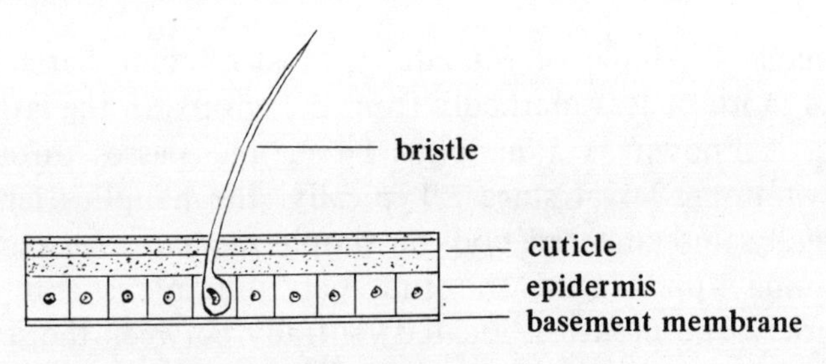

Figure 18. Part of Body Wall showing Bristle

permeable walls are probably receptors of chemical stimuli such as taste and smell. Some crustaceans have balancing organs or statocysts that are sensory pits containing sensory hairs and hard particles such as sand grains. In the decapods they are located at the base of each of the antennular protopods and in the mysids in the endopod of the uropod.

Reproductive System Except for a few parasitic or sessile forms, most crustaceans are unisexual, that is, the sexes are separate. The male organs of reproduction are a pair of glands called testes (Fig. 14), in which the sperms, the male reproductive cells, are formed, and a pair of ducts, the vasa deferentia, through which the sperms pass to the exterior. The female organs are a pair of glands called ovaries, in which the eggs, the female reproductive cells, are formed, and a pair of ducts, the oviducts, through which the eggs pass to the exterior.

In some species of crustaceans, the animals possess both male and female reproductive organs, a condition known as hermaphroditism. It occurs, as a rule in the cirripeds, in some parasitic isopods, and in a few decapods. The production of young from unfertilized eggs, which is known as parthenogenesis, frequently occurs in the branchiopods and ostracods (where it often alternates with sexual reproduction) and occasionally in the terrestrial isopods.

Eggs, Hatching, and Development During the mating process, the eggs are fertilized by the males as the eggs are extruded from the oviducts. Generally, they are carried by the female after extrusion, though they are sometimes shed freely in the water or are deposited on some substratum. When they are retained by the female they may be kept between the valves of the shell, as in the ostracods, in the mantle cavity, as in the cirripeds, in special egg cases that are attached to the genital segment, as in the copepods, in a brood pouch, as in the peracarids, or they may be attached to the abdominal appendages of the female, which is the case in most of the decapods.

Although the eggs in some crustaceans hatch into a form that is a reasonable facsimile of the adults, most of them hatch in a form that differs more or less markedly from the adults. In the latter instance the young is known as a nauplius larva, and passes through a series of free-swimming larval stages. Typically, the nauplius larva has an oval, externally unsegmented body, a simple median eye, and three pairs of swimming appendages, the first pair uniramous and the other two biramous. The mouth is located ventrally between the second and third pairs of appendages and the anus terminally. As the larva grows and development proceeds, the body becomes elongated and the posterior part becomes segmented, new segments being added at each successive molt in front of the anus. Meanwhile, buds appear on the ventral surface of each segment and gradually become differentiated: thus the uniramous appendages become the antennules of the adult, the first biramous appendages become the antennae, and the second biramous the mandibles. Other additional appendages are usually biramous when they first appear. At the same time as the buds appear, the dorsal covering of the body is gradually extended backwards as a shell-fold and paired eyes appear beneath the cuticula in the head region. With further development the larva gradually assumes its adult form. In the species in which the young resemble the adult, the larval stages are passed through in the egg.

Molting Since the integument or exoskeleton is hard and inelastic, it must be shed periodically to allow a crustacean to grow and increase in size. The process is called molting and is not without hazard, since the animal at such a time is without a protective armor and is vulnerable to attack. Before molting a new soft shell is formed under the old hard one.

Regeneration Some crustaceans, such as the crayfish, can regrow or regenerate a lost part as an appendage for instance. The new part, however, is not always quite like the old one.

Autotomy Perhaps the most interesting anatomic structure associated with the regenerative process in the crayfish is the definite breaking point near the bases of the walking legs. If one of them is caught in a crevice or is grasped by a predator or in some manner is injured, the crayfish simply discards it and a new one develops from the end of the remaining stump. This phenomenon is called autotomy and occurs also in a number of other animals.

Autotomy or self-mutilation as it is also known is accomplished by a special muscle called an autotomizer. The leg is flexed by the muscle and by a continued pulling of it the leg is separated at the breaking point. The muscle is not damaged in the process; meanwhile a membrane grows across the inside of the leg on the proximal side of the breaking point and in which there is a small hole through which pass the nerves and blood vessels, but the hole is quickly plugged by a blood clot.

Locomotion Most crustaceans move about by walking or swimming, though some, like the barnacles, are sessile and remain attached to some support as a rock or the hull of a ship.

Size Crustaceans vary in size from the minute forms like the "waterfleas," some of which measure less than a millimeter in length, to the monster crabs that measure up to two feet across their long, spindly, outspread legs.

Food Although the food of the crustaceans may be said to be largely decaying animal and plant substances, many of them also feed on small living animals and plants. Many of them are also parasitic.

Habitats Some crustaceans, such as the sow bugs, land crabs, and a few other forms, live on land, but the crustaceans as a group are essentially aquatic animals, occurring in both fresh and in salt water.

Abundance and Distribution There are said to be some 25,000 species of crustaceans. They may be found throughout the world, are abundant in fresh water, there being hardly a ditch or pond that does not have at least some of the smaller forms, and occur in teeming multitudes in the seas and oceans where a few species penetrate to the greatest known depths. They may even be found in caves and caverns and in the tissues of other animals.

The Fossil Record Fossil remains of crustaceans occur in all the geological strata. Shrimplike forms are present in the Upper Devonian and Carboniferous rocks and decapods in the Triassic.

The History of Crustaceans Excavations in the Near East reveal that crabs were used as food about 20,000 B.C. Doubtless they and other decapods were eaten by man before that time. The Greek philosopher Aristotle was the first to describe them and called them Malacostraca or soft-shelled animals, as distinct from those with a hard shell.

The earlier microscopists observed the lower crustaceans but paid little attention to them, and it was not until the time of the Danish naturalist Müller that they began to be studied. Linnaeus classified them with the *Insecta aptera* or wingless insects. In the early years of the nineteenth century all the crustaceans were classified as Crustacea, a term already in use as a synonym of the Malacostraca, and their grouping into orders was largely the work of the French entomologist Latreille, who introduced the names Branchiopoda, Isopoda, Amphipoda, Decapoda, and Phyllopoda.

The study of American crustaceans can be said to have begun with Thomas Say in the first quarter of the last century. Say and others laid the groundwork for what we know today about the American species.

Luminescence Certain pelagic and deep-sea crustaceans are able to produce light at will. Some of them secrete a luminous substance from dermal glands or from their secretory organs; others have complex light-producing organs called photophores.

Economic Importance of the Crustaceans The crustaceans as a group are of considerable value as food for man, shrimps being the most important commercially, followed by crabs, lobsters, and crayfishes in that order. The crayfishes are used as food in Europe and on the Pacific Coast, and in certain regions the blue or edible crabs are eaten extensively.

In the general economy of Nature the primary role is played by the smaller crustaceans, for they are usually present in enormous numbers in both fresh and salt water where they act as scavengers and serve as food for other animals, including, of course, the fishes. The copepods, for instance, form an important part of the plankton where they feed on diatoms and other minute organisms and in turn are eaten by larger forms.

At times some crustaceans can become pests. Thus in some parts of the Southern states crayfishes often damage cotton and other crops by eating the plants. Occasionally they may burrow into the levees and thus weaken them. Sow bugs, which also feed on vegetation, may become undesirable tenants of greenhouses and in fields can do considerable damage if they become numerous.

Although none of the crustaceans are themselves parasitic on man or other land animals, some of them serve as intermediate hosts for such parasitic worms as the guinea worm and tapeworms.

2 The Decapods

The decapods (from the Latin *deca,* ten; *pod,* foot or leg) are the largest and most highly organized of the crustaceans, and though often grossly different in appearance are, nevertheless, basically similar in functional anatomy. We know them as the lobsters, crabs, shrimps, and crayfishes.

In these animals the thoracic segments are fused with those of the head to form the cephalothorax, which is covered by the carapace, the carapace in some species being more or less cylindrical or compressed sidewise, in others somewhat flattened. They have stalked eyes, a pair of mandibles, two pairs of maxillae, three pairs of maxillipedes, and five pairs of thoracic legs (hence their name of decapods or decapoda), the first being usually much larger than the others and bearing pinching claws. They have usually more than one series of gills and have statocysts in the adults.

The majority of the decapods are marine in habit, though some dwell in fresh water and a few live on land. Their food consists of both live and dead plants and animals, much of it being microscopic in size. The sexes are separate in most species, and the eggs are typically attached to the abdominal appendages (pleopods) of the females until hatching. In some species the newly hatched young resemble the adults in general structure, but in most they differ markedly in form, the form varying in different species. The young or larvae feed on small organisms, molt between each stage, and move about by means of various appendages.

True Lobsters It is not known when the first lobster was eaten by man, nor when it first appeared as a table delicacy. We do know, however, that lobsters were served at a banquet given by Elizabeth of Austria when she made her ceremonial entry into Paris in 1571, so they must have been used as an article of food prior to that time. Writing in his journal of his voyage to the coast of Maine in 1605, Waymouth says that "towards night we drew with a small set of twenty fathoms very nigh the shore; we got about thirty very good and great lobsters . . . which I omit not to report, because it showeth how great a profit the fishing would be. . . . " Today some three and a half centuries later, Maine lobstermen are still finding lobster fishing a rather lucrative business, since they produce about one third of the annual catch. But the days of the "great" lobster are gone, since overfishing and subsequent stringent regulations have prevented all except a very few lobsters from reaching any great size.

Lobsters belong to four different families: true lobsters (Nephropsidae); spiny lobsters (Palinuridae); slipper or Spanish lobsters (Scyllaridea); and deep-sea lobsters (Polychelidae). Some true and spiny lobsters are highly esteemed for their tasty flesh, which has about 215 calories per eight-ounce serving and is high in protein and calcium.

The lobster of the American market is the species *Homarus americanus* (Fig. 19). It occurs from southern Labrador to Cape Hatteras. The average length of the adult lobster caught for the market

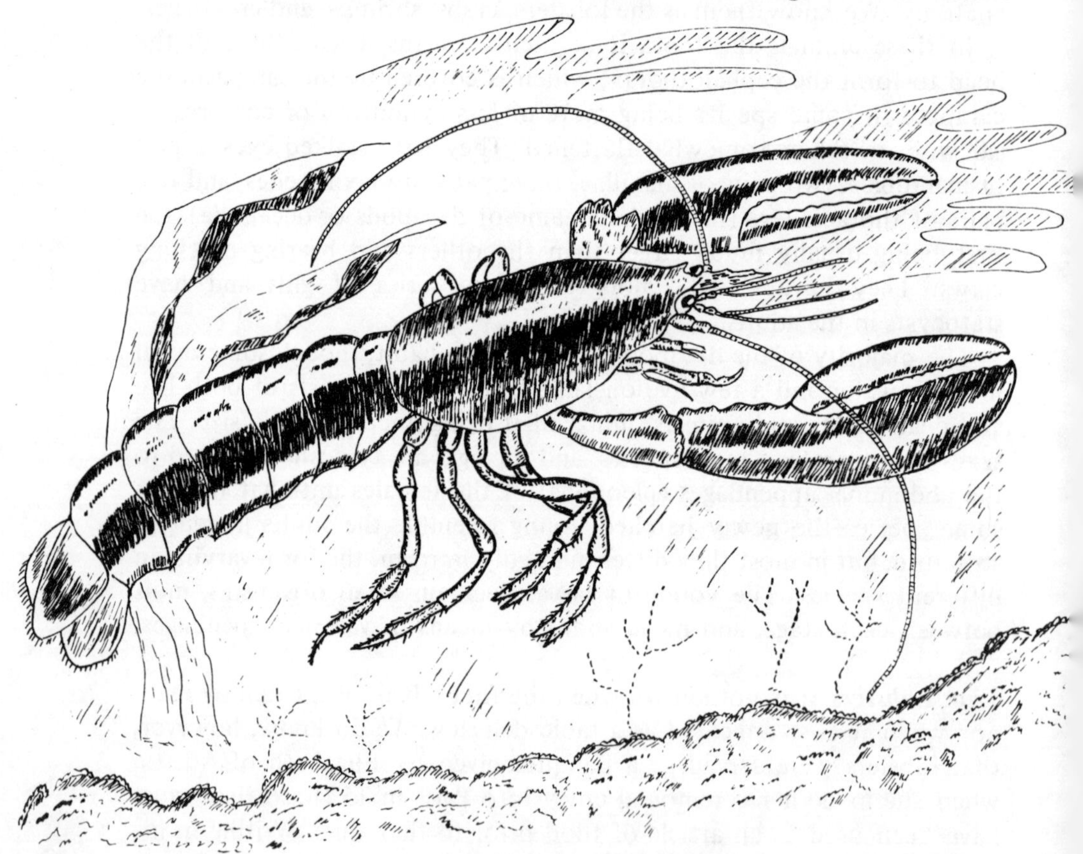

Figure 19. American Lobster

is about nine inches, measured from the tip of the rostrum to the end of the tail, and the weight is from one and a half to two pounds. Occasionally, very large specimens are found, some of them measuring about two feet long and weighing more than thirty pounds. One specimen has been recorded as having a length of thirty-four inches and weighing as much as thirty-five pounds. As far as I know the heaviest lobster ever captured had a weight of forty-four and a half pounds.

Few of us ever get to see a live lobster but most of us know what one looks like. It is a typical arthropod, more specifically a crustacean, but unlike other crustaceans it has a pair of enormously well-developed pinching claws (chelipeds), undoubtedly its most characteristic feature. They are used for both offense and defense.

The body of the lobster consists of twenty-one segments, the first fourteen being united into the cephalothorax, which represents the combined head and thorax, the remaining seven forming the abdomen, and is covered by a hard outer covering or integument, the exoskeleton, which is made rigid from being impregnated with calcium salts. Over the dorsal surface and the sides of the cephalothorax is a single large shield, the carapace. It is folded over the abdomen between the segments to allow for flexibility.

As mentioned in chapter 1, the structural feature of all arthropods is the jointed appendages. If a specimen lobster is carefully examined it will be found that the first segment of the head bears a pair of compound eyes that are set on the ends of jointed, movable stalks. They are not true appendages, since they have a different origin. On the second segment of the head, however, the first pair of true appendages may be observed. They are known as antennules, each of which has two filaments and both of which are sensory in function. Another pair of appendages, called the antennae, are located on the third segment. Unlike the antennules, each has only a single filament. The fourth segment of the head is seen to bear the toothed jaws or mandibles, which serve to crush the food, and the fifth and sixth segments the first and second maxillae that are employed in passing food to the mouth. The second maxilla is a thin, lobed plate and is chiefly respiratory in function, serving as a "bailer" for driving water out of the respiratory cavity.

A glance at the thorax will show that every segment has a pair of appendages. To the first three are attached the first, second, and third maxillipedes, which are somewhat sensory in nature but which are used chiefly in handling food. A gill is attached to both the second and third maxillipedes. The fourth segment of the thorax bears the large claws or pinching legs, the next four each a pair of walking legs. All five pairs of legs have a gill separator and a gill (Fig. 10). As the animal walks, the movements of the legs move the gills and stir up the water in the respiratory cavity under the carapace.

In the lobsters over an inch and a half long, the pinching legs are not symmetrical; in other words they are asymmetrical or not quite alike. In the smallest lobsters they are both slender and have sharp teeth, but as the lobsters grow the legs gradually undergo differentiation, one becoming larger than the other and its sharp teeth fusing into rounded tubercles. This leg is used for crushing. The other leg remains smaller and more slender, its teeth becoming still sharper. It is used primarily for seizing and tearing prey. I might add that the first two pairs of walking legs have small pincers that help in catching prey and that the last pair is used for cleaning the abdominal appendages.

By looking carefully at the abdomen, it will be found that every segment except the last has a pair of appendages. The pair on the first segment differs in the two sexes, in the male being modified to form a troughlike structure (Fig. 20) for transferring sperms to the female during the mating process and in the female being much reduced (Fig. 21). This difference in the appearance of the first pair of appendages

Figure 20. 1st Abdominal Appendage of Male

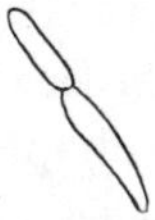

Figure 21. 1st Abdominal Appendage of Female

serves to distinguish the sexes. The appendages on the following four segments are biramous or two-branched and are known as swimmerets (Fig. 13). They are used in swimming forward and in the female for the attachment of eggs. The appendages on the sixth abdominal segment resemble modified and enlarged swimmerets and are known as uropods (Fig. 1). Together with the last flattened segment, called the telson,* they form a tail fan and are used when the animal swims backward.

In the lobster, the abdomen is generally carried extended, but it can be swiftly reflexed or snapped under and forward against the lower surface of the thorax, an action that is important for the swift backward swimming motions that constitute the animal's escape behavior. The large muscles that activate the abdomen are the tissues or flesh we generally eat.

As is the case with many other crustaceans, the lobster has the ability to regenerate a lost or mutilated appendage suffered as a result of fighting or through an accident. However, the new or regrown appendage is usually not as perfect as the original. The lobster also has

*It is debatable whether the telson is a true segment.

the power of casting off a walking leg (autotomy) should such a member be grasped by a predator or be caught in the crevice of rocks or be injured in any way.

The color of the living American lobster is a dark green, mottled with still darker spots, and yellowish or orange on the lower surface. Sometimes there is a bright blue on the limbs and reddish markings may also occasionally be found. When boiled it becomes a bright red, which is the only way most of us ever see it.

Lobsters are marine animals and live on rocky, sandy, or muddy bottoms from the shoreline to beyond the edge of the continental shelf. They dwell singly in crevices or in burrows under rocks. Being nocturnal in behavior, they forage at night, moving about by walking forward on their walking legs or by swimming by means of their caudal fin or more properly the uropods and telson.

Lobsters are fundamentally scavengers, feeding on dead or decaying animal matter. That they can convert such seemingly, at least to us, unappetizing food into the delicious flesh that has become a table delicacy appears nothing short of the miraculous, on a par perhaps or even exceeding the remarkable wonders that chemists can perform in the laboratory where they turn all sorts of chemicals into the many useful products with which we are familiar.

But lobsters are not entirely scavengers; they also eat live fish, dig for clams, and feed on algae and eel grass. They have at times been seen to attack large gasteropods, breaking off the shell piece by piece to get at the soft inner parts. Their food is shredded by the maxillipeds and maxillae, the jaws further crushing it before it is conveyed to the mouth. From the mouth the food passes to the stomach, part of which is specialized as a gizzard; the latter is lined with hard, chitinous teeth and is operated by numerous sets of muscles. In the stomach the food is pulverized, strained, and sorted, the smallest particles being sent in a fluid stream to the large digestive glands where they are digested and absorbed, the coarsest particles being returned to the gizzard for further grinding. Particles of food that cannot be digested are passed to the intestine and ejected through the anus as feces. As an afterthought we should also add that lobsters are somewhat cannibalistic, preying on their weaker brothers and sisters. If the young lobsters were not scattered by the ocean currents, lobsters would be in jeopardy of eating themselves into extinction.

The circulatory system of the lobster is an open one, that is, the blood, which is pumped by a muscular heart situated dorsally in a blood-filled chamber called the pericardial sinus, flows freely throughout the body and into blood cavities called sinuses where it bathes the tissues, reentering the heart through openings or ostia.

The excretory organs, which are sometimes called green glands because of their greenish color, consist each of a glandular sac and a coiled tube that opens into a muscular bladder. The waste products of metabolism are extracted from the blood and passed into the bladder from which they are ejected to the outside by means of a pore at the base of each antenna.

In order to supply the demands of such a large and rather active animal as the lobster, there is an extensive respiratory surface that is provided by twenty pairs of gills. They are attached to the bases of the legs, the membranes between the legs, and the wall of the thorax and lie on each side of the body in a cavity enclosed by the curving sides of the carapace. Water enters the cavity, passes upward and over the gills, and then out anteriorly in a current produced by the flattened plates of the second maxillae.

The nervous system of the lobster consists of a mass of nerve cells, the brain, which is located in the head near the eyes, and a ventral nerve cord that extends backward and has a pair of ganglia in almost every segment and to which the brain is connected by a pair of connectives that pass around the food tube. Since the lobster is more or less a nocturnal animal and, moreover, lives at depths where there is little if any light even during the day for clear vision, the eyes are probably of less importance as sense organs than the sensory bristles that are distributed all over the surface of the antennae, body, and appendages. There are from fifty thousand to one hundred thousand of these bristles on the pincers and walking legs alone and they are of two kinds–one sensitive to touch and the other to chemicals. It is undoubtedly these bristles that enable the lobster to detect changes in its environment and so be able to adjust itself accordingly.

On the basal segment of each antennule there is a water-filled sac that opens to the outside by means of a fine pore, and on the floor of the sac there is a ridge of sensory fine hairs along which are distributed many fine sand grains. This structure appears to be an organ of equilibrium or statocyst, for if the sand grains are removed the animal becomes disoriented or perhaps we should say loses its powers of orientation.

Since the outer integument of the lobster is hard and cannot stretch, the lobster cannot grow unless it periodically gets rid of it, a process, as we have seen in chapter 1, called molting. Molting usually occurs in summer or fall. The animal molts eight times the first year, five times the second, and three times the third year, after which the male molts twice and the female once a year.

At the time of molting, the lobster retires to some secluded spot, for molting is always attended with many dangers, and without its protective armor it is vulnerable to attack. During molting the lobster lies on its side, the back splits open longitudinally, and the animal then gradually works its way out of the old shell. It is rather amazing how the lobster can free its fleshy claws, which are as much as four times the diameter of the joints through which they must pass. To effect this seemingly impossible feat, blood flows out of the claws prior to molting and causes them to shrivel. Then the lime in the joints dissolves, which thus softens the joints and the lobster is thereby able to draw its shrunken limbs through the small passages. That the claws and whole limbs are sometimes lost during the molting process is not entirely unexpected; however, the lobster is able to replace them. At first the replaced limbs are small but they gradually grow to somewhat the normal size during the next several molts. Molting takes about five to twenty minutes, but it takes considerably longer for the new shell to fully harden, some six to eight weeks.

As winter approaches, the American lobster moves from shallow rocky bottoms toward deeper water, where the temperature is more comfortable and where food is attainable, and then mates. Mating usually takes place a few hours after the female has molted, while her shell is still soft, the male having molted several weeks earlier. During the mating process, the male turns the female on her back and then transfers the sperms, which have passed down a pair of ducts leading from the testes, in which they were produced, and through a pair of external openings at the bases of the fifth pair of legs, to the female's sperm receptacle, which lies between the bases of her last two pairs of legs. They remain alive in this receptacle and fertilize the eggs, formed in the ovaries, when they are released through the oviducts the next time the female molts.

At the time that the eggs are released, which may take several hours, the female lies on her back and flexes her abdomen. The tiny round eggs leave the oviducts through external openings at the bases of the third pair of legs and then pass over the sperm receptacle where they are fertilized. Once fertilized, the female transfers the eggs to the swimmerets and attaches them by means of a sticky secretion. The average female lays between eight thousand and ten thousand eggs.* Once attached to the swimmerets, the eggs are incubated for several months and kept aerated and clean by movements of the swimmerets. The eggs, or "berries" as the lobstermen call them, require from ten to eleven months to hatch. After the young lobsters have emerged from the eggs, they cling to the mother for a time and then are dispersed by a vigorous shaking of the swimmerets.

*The largest number of eggs ever found on a female was 97,400.

The young lobsters are free-swimming and differ from the adults in appearance (Fig. 22). The appendages are all biramous, similar structures, but the swimmerets during the first larval stage, when the young lobsters are only about one-third of an inch long, are merely small buds, the young losbters swimming about at the surface by the rowing action of the flattened, fringed outer branches of the limbs. Young lobsters pass through three larval stages in about twelve days. During the fourth stage (Fig. 23), when they are about three-quarters of an inch long and resemble their parents and then swim forward by means of their swimmerets, they seek a suitable place in which to settle. On finding such a place they molt into the fifth stage and spend the remainder of their lives on the bottom unless caught by a lobsterman or eaten by a predator.

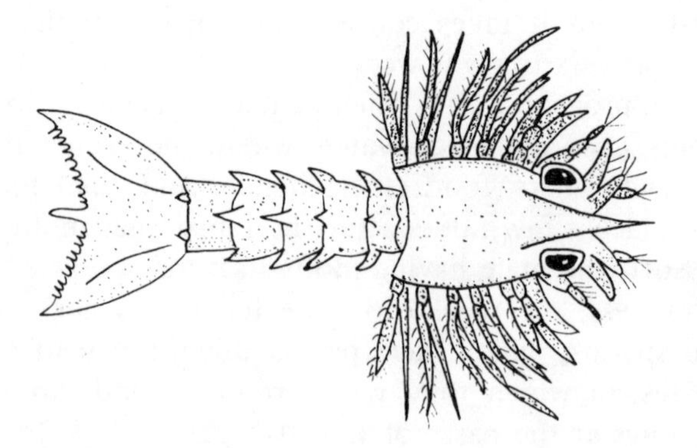

Figure 22. First Larval Stage of Lobster

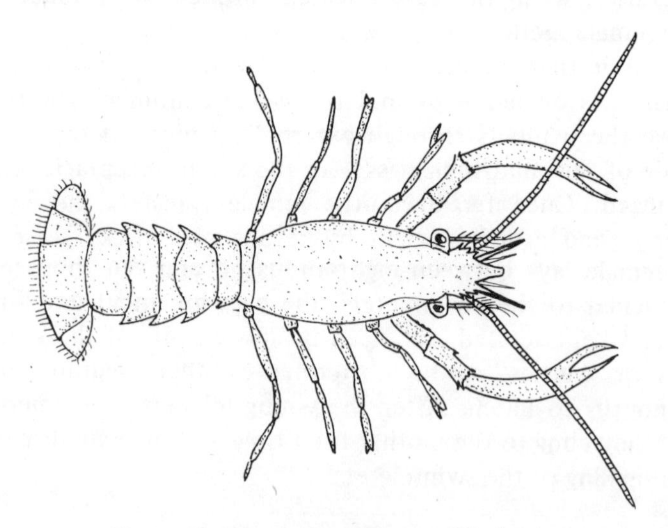

Figure 23. Fourth Larval Stage of Lobster

Female lobsters produce prodigious numbers of offspring simply to ensure that enough will survive to perpetuate the species. Small lobsters are just the right size to provide many small predators with a meal or larger ones with tasty tidbits. Tautog, skate, and dogfish take a heavy toll and cod are especially fond of them, next to man being their most deadly enemies.

Lobsters are generally fished in relatively shallow water, five to fifty fathoms deep, and are caught in traps called lobster pots. The traps are made of wood laths and have a flat floor and a rounded roof. At each end there is a funnel of netting or cord, through which the lobsters enter. They are lured to the traps with clams or fish as bait and once inside the traps are unable to get out. Lobsters are generally shipped and marketed alive.

As with many of our other natural resources no thought was given at one time to conserving or developing the supply of lobsters. Not too long ago the annual catch for Canada and the United States amounted to over one hundred million. Today it is a fraction of that number. To save the lobster industry from itself, most states now have stringent laws prohibiting the sale of small-sized lobsters.

There are three species belonging to the genus Homarus, *H. americanus,* the American lobster, *H. gammarus,* the European lobster, and *H. capensis,* and there are about forty species of true lobsters. The European lobster, which ranges from Norway to the Mediterranean Sea, is similar in habits to the American species and like it is also of considerable commercial importance. It reaches a weight of fifteen pounds, though two- to three-pound specimens are most common. *H. capensis* found about the Cape of Good Hope has no commercial value.

Also commercially important is the Norway lobster, *Nephrops novegicus.* It is a slender species with slim claws. The males may be twelve inches long, the female nine inches, but most specimens that are caught measure only five to six inches. Also known as the Dublin lobster and the Italian scampa, its range is from Iceland and northern Norway, Scotland, and Ireland to the Mediterranean Sea.

Spiny Lobsters The spiny lobsters belong to the family Palinuridae and may be distinguished from the true lobsters by the presence of spines on the carapace and the absence of claws on their legs. Their abdomen is well developed, extended, except when it is used to propel the animals vigorously backward, symmetrical in shape, and covered with a spiny armature. The telson and the last pair of uropods form a broad tail fan (Fig. 24).

Unlike the true lobsters, the females lack a sperm receptacle; instead the sperms, enclosed in capsules, called spermatophores, and which are

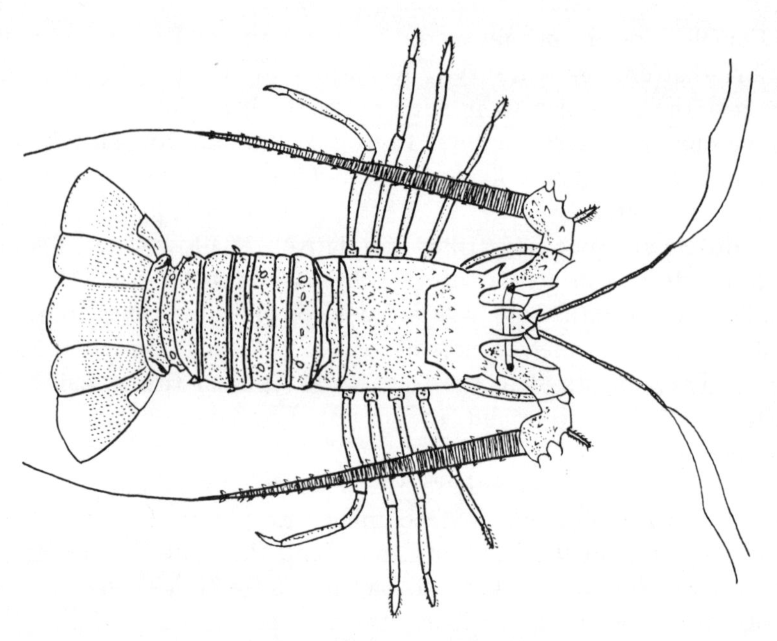

Figure 24. Spiny Lobster or Sea Crawfish *(Panulirus argus)*

embedded in a puttylike matrix, are deposited on the underside of the female's thorax. The matrix is white and soft at first but darkens and becomes a hard sperm case. When the female later releases her eggs she breaks open the sperm case.

The spiny lobsters have an unusual larval form, called the phyllosoma, which is not found among other crustaceans. It consists of a very thin, flat, transparent, roughly circular disk with eyes and legs projecting from the margin.

Although the spiny lobsters prefer coral reefs or rocky situations, they may also be found over smoother sandy bottoms, if sheltered niches or brush piles are available. They avoid muddy grounds, however. Their food habits are similar to those of the true lobsters and they consume considerable invertebrate food. Like the true lobsters, they also forage at night.

There are about forty-five species of spiny lobsters, several of which are of major economic importance. They may be found throughout the world, chiefly in tropical and subtropical latitudes, though they also occur in certain temperate regions.

Quite common on the coral reefs and under rocks in coral lagoons from North Carolina to Florida is the species known as the West Indian Spiny Lobster *(Panulirus argus).* They measure eight to ten inches in length with variegated colors, their patterns ranging through dark brown, red, and yellow with bluish areas. They are an important food animal. This species and species imported from southern Africa and Australia form the source of the "lobster tails" sold in the stores and restaurants of the United States.

A somewhat similar species is the California spiny lobster *(P. interruptus).* It is often common in tide pools and in shallow water along the southern California coast. A European species *(P. elephas)* occurs from Great Britain and France to the Mediterranean Sea. Probably the most widely distributed species is *Janus lalandii.* It is found along the coasts of western and southern Africa, Australia, New Zealand, Tasmania, and about the islands of Juan Fernandez (off Chile), Tristan da Cunha (south Atlantic), and St. Paul (south Indian Ocean).

Slipper Lobsters There are about seventy species of slipper lobsters. They belong to the family Scyllaridae and are characterized by having a broad and flat body, short and scalelike antennae, and eyes in sockets in the carapace. They measure from ten to fourteen inches in length and weigh as much as five pounds. All are marine and live in shallow water, buried in mud or sand, and occur in warm waters throughout the world. The slipper lobsters have a larval form like that of the spiny lobsters.

Deep Sea Lobsters These lobsters belong to the family Polychelidae and number about seventy species. They are all blind and live at great depths, buried in muddy bottoms. Their integument is soft and flexible, and the first four and sometimes all five pairs of legs end in claws. They feed mainly on dead organisms that drift down. Surprisingly perhaps, fossils of these animals were known long before living specimens were found. These fossil forms lived in shallow water and had functional eyes.

True Crabs The true crabs are the highest of the crustaceans. Unlike the lobsters, who have a long and cylindrical body and whose abdomen is extended, the crabs have a flat and broad body and a short abdomen that is bent sharply under the thorax where it fits so perfectly in a groove that its dorsal side is flush with the ventral side of the thorax, thus becoming quite hidden from above. In the crabs the eye stalks are long and fit into sockets on the carapace, both the anntenules and antennae are small, the antennules frequently folded into small grooves, and the third pair of maxillipeds are broad and flat, rather platelike, and cover the mouthparts like a lid. The first pair of walking legs are comparatively large and end in pinching claws; the other four pairs end in simple points except in the swimming species when the fifth pair is flattened to form fins or swimming paddles. The abdominal appendages or pleopods are few in number: the male has only two pairs, which are modified as copulatory organs; the female has only four pairs, which she uses for carrying her eggs.

The life history of the true crabs is quite complicated, and there are several stages of somewhat singular metamorphoses before they become adults. The most marked forms are what are called the zoea and the megalops, and these are so unlike the adults that at one time they were classed as distinct genera. After the young or larva has molted several times it has an appearance such as is shown in Figure 25, which is the last zoea stage. From this stage it changes to the megalops stage (Fig. 26). The megalops has enormous eyes, an extended abdomen, an

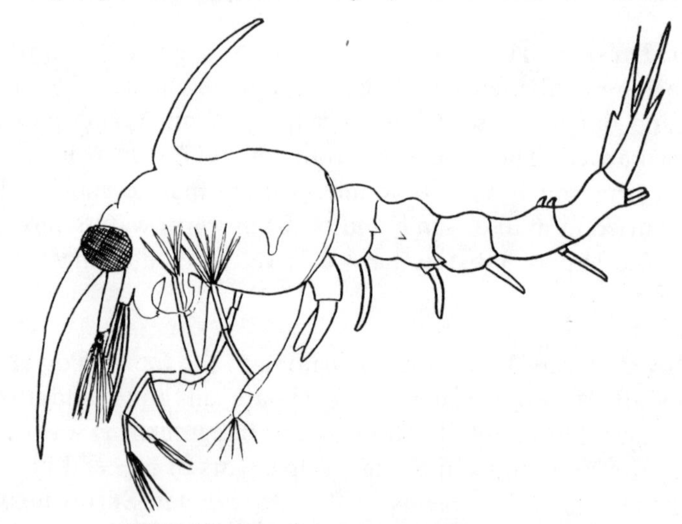

Figure 25. Last Zoea Stage of Crab

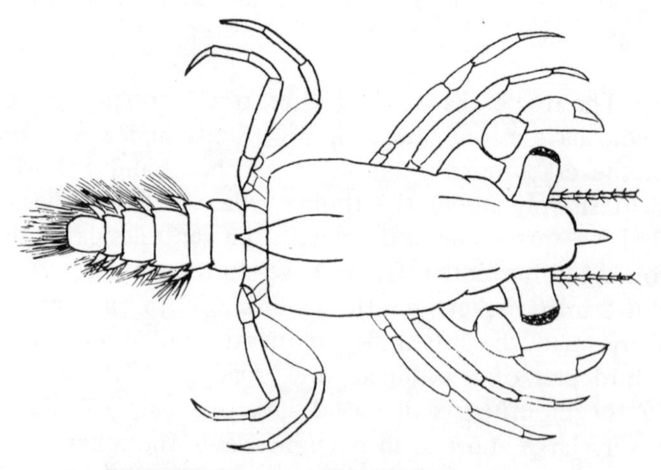

Figure 26. Megalops Stage of Crab

elongated carapace, and swimming legs. This stage is a brief one, and at the first molting the larva changes to a form that is rather similar to that of the adult. From now on the larva grows by shedding its shell at certain periods until it has reached its full growth, after which it is not likely that it molts again.

There are some one thousand species of true crabs ranging in size from that of a wheat grain to fifteen inches or more across the back. They occur mostly in the sea where they live on or near the bottom, from tide marks to very great depths. Some, however, swim so well and so rapidly that they are frequently seen near the surface. Some species live near the high-water mark or above it and have become more or less terrestrial animals; these run rapidly over the sand in which they dig deep burrows, their gills having been modified to absorb and retain the moisture in the damp sand. Still others live on land and are habitually found far from water to which they return periodically to deposit their eggs. A few live exclusively in fresh water. Crabs are found all over the world.

Not only is there a diversity in habitats, there is also a diversity in habits. Thus they creep, climb, swim, and burrow, in structure being modified to their own peculiar mode of living. There is also considerable variation in their shape as well as in their color and markings. Like other crustaceans they are for the most part scavengers, but the land forms also include plant substances in their diet. Crabs are great fighters and very wily, often resorting to a stratagem to avert danger. They have a good deal of intelligence and many amusing habits; in short they are not only a curious group but an interesting one as well.

True Crabs–Swimming Crabs The swimming crabs (family Portunidae) have a transversely oval body with the last pair of legs more or less adapted for swimming, being broad and flattened to form effective paddles. One of the more interesting of these crabs is the green crab *(Carcinides maenas),* quite common on the Atlantic coast north of New Jersey, where it may be found between tide marks, often quite some distance up on the beach where it hides beneath stones. It may also be found in tide pools and in holes and cavernous places on the shore. It is an attractive little crab, from one and half to two inches long and a little more in width, with five acute teeth on each side of the anterior part of the carapace, and in color is green, spotted with yellow, making it rather conspicuous. Unlike other swimming crabs whose last pair of legs are expanded as oval paddles, those of the green crab end instead in pointed tips (Fig. 27).

The green crab, which is cosmopolitan in distribution and in Europe

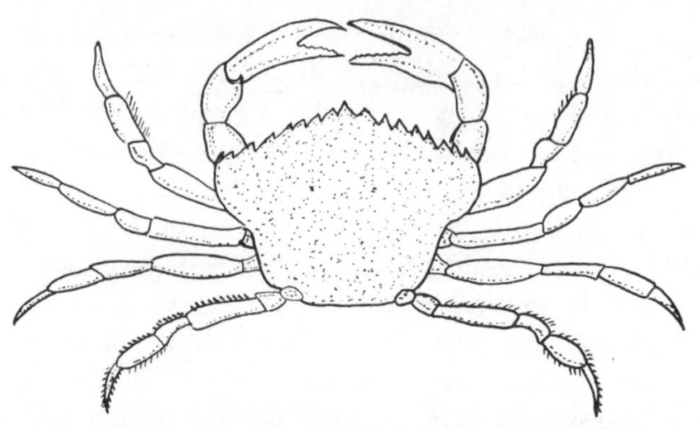

Figure 27. Green Crab *(Carcinides maenas)*

is used as food, is a lively creature, and I have often been amused at its behavior as I have watched it run about on the beach. When cornered it displays a reckless audacity, hence its specific name. The French have a good name for it. They call it *crabe enragé.*

A second species of swimming crab that is quite common on the sandy shores from Cape Cod to Florida is the lady crab or calico crab *(Ovalipes ocellatus).* It occurs among the loose sands at low-water mark and also on sandy bottoms offshore. At the low-water mark it has the habit of burying in the sand up to its eyes and antennae and in such a position watches for both prey and foe. Should danger threaten at anytime, it quickly disappears beneath the sand. By burying deeply in the sand it escapes the action of the breakers.

The general shape of the lady crab (Fig. 28) is circular, with the carapace about as long as it is broad. There are five conspicuous teeth on the margin of either side, and on the front margin there is an indentation on each side of a three-spined rostrum to form cavities for the eyes. The first pair of legs are large and have claws, the last pair are flattened into oval swimming paddles, and the intermediate three pairs are simple in structure and end in points. In color the carapace is white or light lavendar and is covered with purple or red spots. In the southern states the lady crab is used for food.

The third species of swimming crab is the blue crab *(Callinectes sapidus).* It occurs along the Atlantic Coast from Cape Cod south to Florida and around the Gulf of Mexico to the Mississippi, and next to the lobster is our most valuable food crustacean being highly valued for its edible qualities. It is also known as the hard-shell or soft-shell crab, suitable for the market not only when the shell is hard but also immediately after molting before the new shell has hardened.

The carapace is about as twice as wide as it is long, averaging about six inches, and has a long, sharp projecting spine on each side, which is

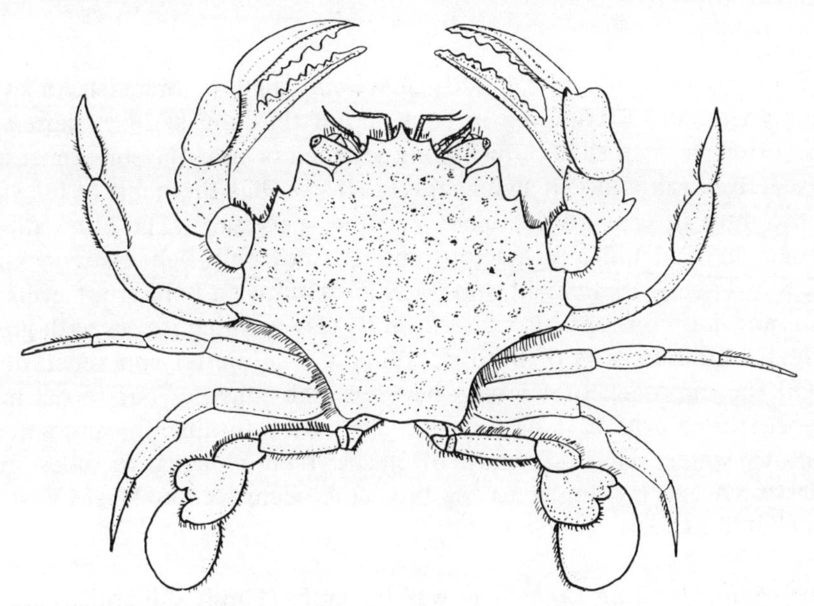

Figure 28. Lady Crab ***(Ovalipes ocellatus)***

a distinguishing feature. There are also eight short spines on each side and between the eyes four unequal teeth and a small spine underneath. The first pair of legs are large and somewhat unequal in size, the last pair are modified into swimming organs, and the intermediate legs are simple in structure. The upper surface of the body is dark green, the lower surface is dingy white, the legs blue, sometimes suffused with red, and the tips of the spines reddish (Fig. 29).

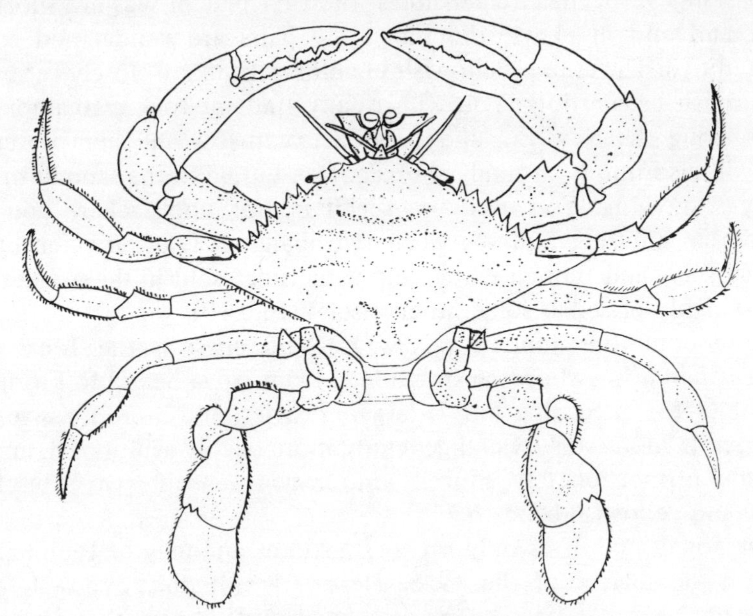

Figure 29. Blue Crab ***(Callinectes sapidus)***

The blue crab inhabits muddy shores, in shallow, brackish, or even fresh water, and is common in bays and at the mouths of estuaries. It may often be seen swimming among seaweed or near the surface. It is a very active crab and can swim rapidly. It also has the habit of burying in the mud in case concealment becomes necessary. The blue crab is predaceous and quite pugnacious and will not only fight its own kind but any enemy that might approach it. It shows a bold front even to man, and has considerable strength in its claws, which it uses with great skill. Its food consists of both decaying animal matter and vegetation.

On the approach of winter, the blue crab migrates out to sea into deeper water where it can escape the ice, returning in summer to shallow water where it may often be seen clinging to piles and wharves. A species similar to the blue crab occurs on the West Coast of the United States.

True Crabs—Walking Crabs The walking crabs (family Cancridae) are so named because the last pair of legs are adapted for walking. They have a rather broadly oval carapace, usually wider than long with a very short rostrum or none at all. The anterior margin is arched and toothed and the last pair of legs are pointed at the end. A common species along the New England coast is the rock crab *(Cancer irroratus),* whose range is from Labrador to South Carolina, though it is rather rare south of New Jersey. The carapace is convex with nine blunt teeth along each side of the front edge and is covered with fine granulations. The eyes are set on short stalks in deep, circular holes, the first pair of legs are short and stout and end in claws, the remaining pairs are slender and end in pointed tips. The crab is yellowish in color with many closely set brown or purplish brown dots. You will usually find the rock crab among the rocks along the New England coast between low and high water but you will also find it on sandy shores, often buried in the sand with only its eyes showing. You may also see it in tide pools where you may witness an amusing fight between two males. It is often eaten by the larger fishes, and though edible and sometimes sold in the market, it is not so highly prized as some other species (Fig. 30).

Quite common along the New England coast is the Jonah crab *(Cancer borealis),* whose range extends from Nova Scotia to Florida. It resembles the preceding but is larger, heavier, and more massive. The carapace is more convex and generally more deeply sculptured, and the legs are proportionately shorter and heavier. In color it is brick-red above and yellowish beneath.

The Jonah crab lives only on rocky shores and may be found at low tide on the surface of the rocks. Here it is not only exposed to the waves but also is subject to the attacks of birds of prey that feed on it.

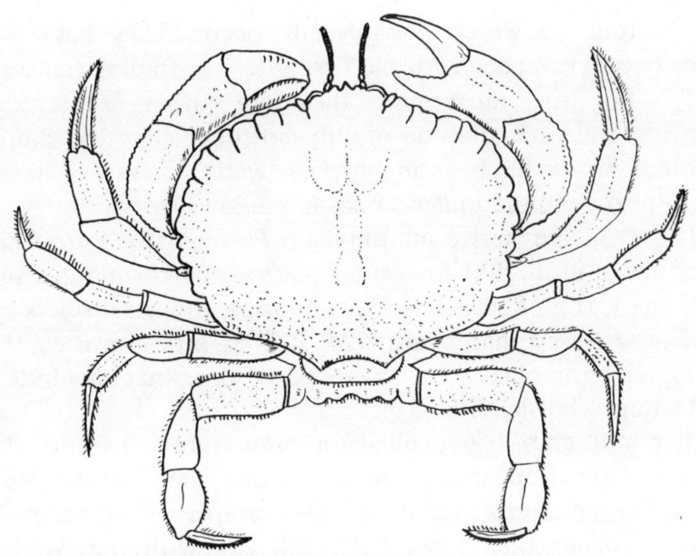

Figure 30. Rock Crab *(Cancer irroratus)*

Irroratus escapes their depredations by remaining concealed beneath the rocks or by lying buried in the sand.

Several species of walking crabs occur along the Pacific coast. *C. magister,* which is the edible crab of California, has a carapace whose anterior margin with nine small teeth on each side is an almost regular elliptical curve. At the end of the curve a large, pointed tooth projects directly outward, and from this tooth the carapace slopes abruptly backward forming a narrow posterior end. It is light reddish-brown, shading to a lighter color in the back; the legs and the under surface of the crab are yellowish. This species may be found in sandy bottoms below tide mark.

Another crab of the family found along the Pacific coast is the species generally called the red crab *(C. productus).* The carapace is somewhat elliptical in outline with five teeth on the anterior margin. In color the crab is dark red above, whence its name, and yellowish beneath as an adult; in the young the color is variable, sometimes being yellow spotted with red, or banded with red and yellow. The red crab lives on rocky shores.

A third species found along the Pacific coast is the rock crab *(C. antennarius),* which inhabits rocky bottoms below low-water mark. It is dark reddish brown above and blotched with red beneath. It may easily be recognized by its hairy antennae, the hairy margins of its abdomen, and walking feet, and the many hairs on the lower surface of the body.

True Crabs–Mud Crabs The mud crabs (family Xanthidae) are small, dull-colored animals of a somewhat muddy hue that blends with the muddy bottom on which they usually occur. They have a usually transversely oval carapace with the first antennae folding transversely or obliquely. For the most part they are rather stout, somewhat slow-moving crabs and live on muddy bottoms along the shore, often being found on oyster beds in brackish water or even in fresh water. They may also be found under stones in muddy places.

The largest of our native mud crabs is *Panopeus herbstii,* which may be found living in mud at low-water mark or burrowing in banks near high-tide mark from Massachusetts to Florida, though it is less common north of New Jersey than southward. It is dark olive brown, the claws being tipped with black. The carapace is somewhat quadrate with a toothed anterior border (Fig. 31).

Another mud crab, often called the mud crab, and found along the coast from Cape Cod southward and along the Gulf Of Mexico to Texas, is *Eurypanopeus depressus.* The carapace is transversely oval, somewhat broader than long, flattened, and with the front margin

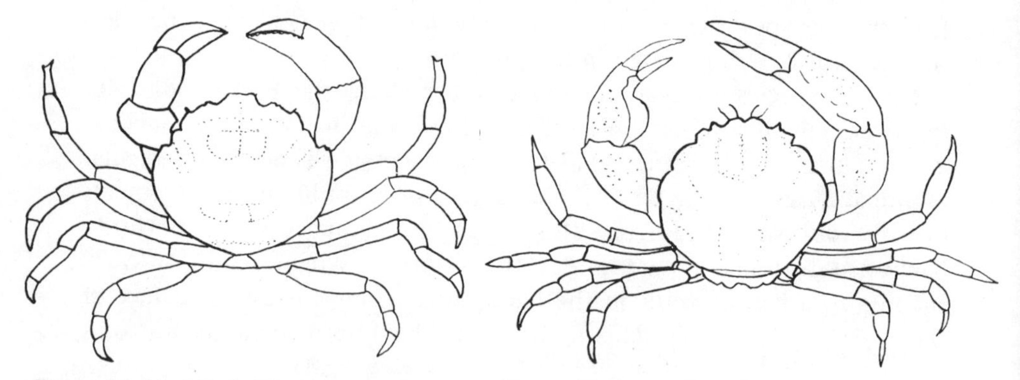

Figure 31. Mud Crab *(Panopeus herbstii)* **Figure 32. Stone Crab *(Menippe mercenaria)***

nearly straight. A third species is *Neopanopeus texana.* It is somewhat convex above. A fourth species easily recognized from other mud crabs because its claws are not tipped with black is *Rhithropanopeus harisii.* It lives near high-water mark and also in salt marshes and occurs along the entire Atlantic coast.

One of the best known of the mud crabs, largely because it is used as food in some localities, is the stone crab *(Menippe mercenaria)* (Fig. 32). It measures about three by four and a half inches and from one to two inches thick. It is a stoutly built crab with a very hard shell, whence its name. The carapace is transversely oval and one of the first pair of legs is larger than the other. Both legs, as a matter of fact, are enormously large and both are tipped with black. The terminal joints of the other four pairs of legs are thickly fringed with hairs and terminate in naillike ends. In color the crab is dark purplish blue, the legs with red and yellow bands.

The stone crab, found from North Carolina to Mexico, lives in deep holes in the mud along the borders of creeks and estuaries as well as in crevices between the rocks of a stone heap. It is a powerful and slow-moving animal but is not pugnacious. One manner of capturing it is to thrust the hand and arm into a hole occupied by it and then drag it out, a rather hazardous method, since one can be badly pinched. The crab is removed with difficulty, since it offers considerable resistance, bracing its claws against the sides of the hole, or if found in a crevice of a stone heap holding on to the rocks with such tenacity that it is often torn apart. A better method and one that is more usually employed is to thrust a hooked iron rod into the hole; the crab seizes it and then may quickly be withdrawn.

True Crabs–Fiddler Crabs In the fiddler crabs (family Ocypodiae), one of the chelipeds or pinching claws, usually the right, of the male is enormously developed, being considerably larger than the other, which is comparatively small. It is a distinguishing feature that serves to identify them without any further description. The large claw is carried horizontally in front of the body and has been likened to a fiddle, the smaller one to a bow, and this together with the waving motion of the large claw has earned them the popular name by which they are generally known. The carapace of these crabs is more or less square, and the eyestalks are usually very long, each lying horizontally in an elongated groove along the front of the carapace. The claws of the female are small and of equal size. The color of the fiddler crabs is a light brown, mottled with purple and dark brown, a color scheme that harmonizes well with the dark sand of the salt marshes and the mud and sand flats where they live. In such places they congregate in immense numbers and excavate burrows, which serve as a home, in convenient places above the reach of the tides–on salt marshes, far up the estuaries and along the mouths of rivers, often where the water is quite fresh. The burrows are an inch to two inches in diameter and often a foot or two deep.

One species of fiddler crab, *Uca minax* (Fig. 33), builds an observation house over the mouth of its burrow in which it sits and surveys its surroundings but quickly retreats into its burrow should danger threaten. The crab excavates its burrow by scraping up the sand or mud and forming it into pellets. These it carries to the mouth of its burrow, but before emerging it peers cautiously about. Satisfied that no danger threatens, it emerges and carries its load four or five feet away before dropping it. It again looks about before returning to its burrow,

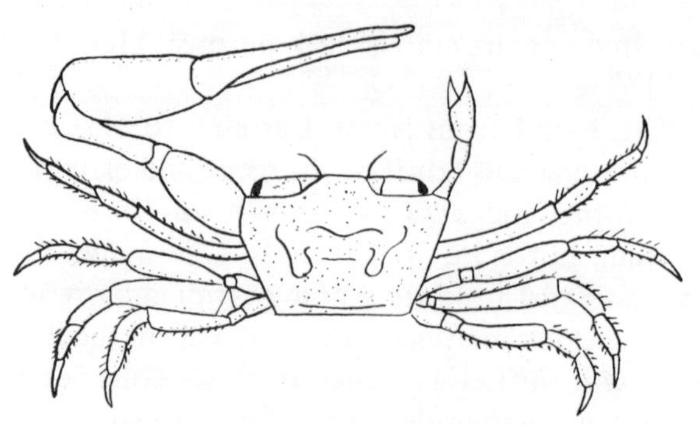

Figure 33. Fiddler Crab *(Uca minax)*

into which it quickly disappears if unable to detect any lurking enemy, soon to return with another load. The burrows cover large areas, and the crabs are usually so abundant that the marshes and shores seem to be alive with them. When frightened, the males lift their large claws and both males and females seek their burrows, running quickly and sideways, after the manner of all crabs, to escape any real or fancied danger. As many retreat into the most convenient burrow, the real owner is apt to find it occupied, whereupon it unceremoniously proceeds to pull out the intruder.

It is rather amusing to watch the male fiddlers sitting at the mouths of their burrows waving their large claws back and forth. It might seem as a friendly gesture, but actually the males are trying to attract a passing female. Should she respond she is led into the burrow with elaborate gestures.

Uca minax occurs from southern New England to Florida and may be found on salt marshes farther away from the sea than other species, and often where the water is quite fresh. It is the largest of our fiddlers and may be distinguished from the others by a red patch at the joints of the legs. *Minax* is essentially vegetarian in its diet, living on small algae. It is capable of living out of water, and without food, for several days.

Two other species may be found along the coast from Cape Cod to Florida: *Uca pugnax* and *Uca pugilator. Pugnax* lives on muddy banks and ditches of salt marshes, the banks frequently being honeycombed by this species, which is exceedingly abundant; *pugilator* occurs on sandy and muddy flats and beaches near high-water mark. On the Pacific Coast, *Uca musica* ranges from Vancouver Island to Mexico and *Uca crenulata* along the coast of southern California, south of San Diego. There are many species of fiddler crabs found all over the world, usually about the same color as the sand and mud on which they live.

True Crabs–Sand or Ghost Crabs Also members of the same family as the fiddler crabs are the sand or ghost crabs. They are square-bodied, long-legged crabs that live in holes in the sand along the seashore and run very swiftly, so swiftly in fact that if you try to catch one it is likely to outdistance you. They are also very dexterous in burrowing and inhabit holes, often three feet deep, which they dig perpendicularly in the sand.

These crabs wander quite some distance from their burrows when the tide is out, and every now and then they extend their stalked eyes and stand on tiptoe to survey their surroundings. When alarmed they hurriedly seek the nearest burrow, but if threatened danger is near they press themselves on the sand only to spring up and dart away if an attempt is made to touch them. When running they hold their bodies high and dodge about so it is difficult to catch them.

A familiar inhabitant of our eastern sandy beaches from Long Island southward is the species *Ocypoda albicans,* generally known as the sand or ghost crab. The generic name means "swift-footed," which certainly applies to this species which is noted for its rapidity of movement, running rapidly sidewise over the sand on the tips of its toes. It is pale blue in color and blends so well with the white sand on a cloudy afternoon or evening that should it suddenly stand still it appears to have completely disappeared, hence its name of ghost crab. It avoids sunny weather whenever possible, since its shadow betrays it as do the odd black eyes that are set on stalks.

The sand crab inhabits sandy beaches above the tide mark and is partial to sand dunes; you can often see hundreds of burrow opening on the seaside face of the dunes. It has become a completely terrestrial animal. If you walk along a beach inhabited by only a few sand or ghost crabs you will likely not be aware of them unless you are very sharp-eyed, but come upon a large colony and your attention will be then directed to them by their numerous dodgings and you will be able to pick out individual crabs scurrying ahead of you, only to disappear as one by one they enter their burrows.

Figure 34. Ghost Crab ***(Ocypode albicans)***

The carapace of the sand crab is almost square in outline with a spine on each anterior corner and large grooves for the eyestalks. One claw is a little longer than the other, and both are coarsely granulated. The remaining eight legs are thickly fringed with hairs (Fig. 34).

The sand crab is more or less nocturnal in habit, remaining for the most part in its burrow during the day but coming out at night to hunt for food, subsisting largely on beach fleas, which it springs upon in the manner of a cat catching a mouse.

There are three American species.

True Crabs–Spider Crabs The carapace of the spider crabs is more or less round in form, rather thick, and tapers in front to a prominent rostrum. The surface of the body is generally rough and irregular with tubercles, spines, prickles, and hairs. Their legs are long and gangling. Some species have the curious habit of drawing their eyes back into the head and thus concealing them. Some, too, have the habit of placing on their backs seaweeds, sponges, hydroids, and other organisms, presumably to hide them from their enemies.

The spider crabs (family Maiidae) are all marine and littoral and are found along both the American and European coasts. They are sluggish creatures and crawl about on the bottom harbors and bays where they are a nuisance to lobstermen and fishermen in general, since they enter the traps and eat the bait.

A common species along the Atlantic coast is *Libinia emarginata,* which is generally known as the common spider crab. It is found on muddy shores and flats, among decaying seaweed, and in eel-grass. It is even found beneath the surface of the mud. The carapace is evenly rounded with a median row of nine spines. Its long legs and sac-shaped body reminds us of a huge spider (Fig. 35). It is brown in color. *Libinia*

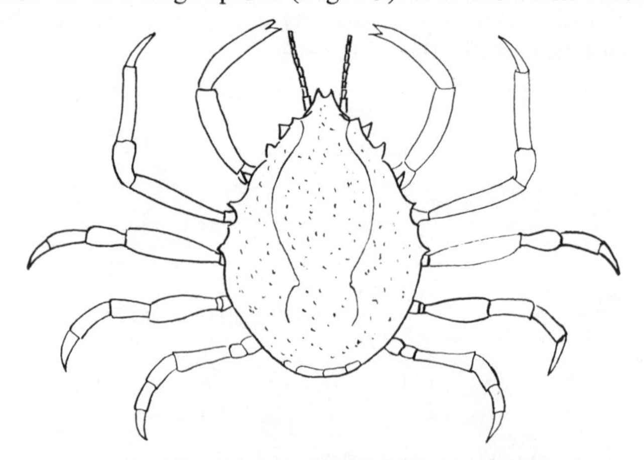

Figure 35. Spider Crab *(Libinia emarginata)*

dubia is similar to the preceding in habits and form but may be distinguished from it by having only six spines in the median row. It is more commonly found in shallow water near the shore and unlike *emarginata* does not occur north of Cape Cod.

A third member of the family but of a different genus is *Hyas coarctatus,* known as the toad crab (Fig. 36) because its body, both in form and size, resembles a toad. It occurs on muddy or stony bottoms from Greenland to New Jersey and rarely southward but then only in deeper water. A related species, *Hyas lyratus,* called the lyre crab because of its lyre-shaped carapace, is dull pinkish red, and is found from Puget Sound to the Bering Sea.

Figure 36. Toad Crab *(Hyas coarctatus)*

The spider crabs seem to select instinctively such seaweeds, sponges, and other organisms that will stand transplanting and also organisms that will afford them maximum concealment. Thus a Hyas, which was covered with bright-colored algae, when transferred to a locality where only sponges grew, was seen to remove the algae and to replace them with the latter. The spider crabs use their chelipeds in transferring various organisms to their backs but before doing so they moisten them with a secretion of mucus or cement from their mouths. At times bivalves and barnacles affix themselves to the crabs, which then have to carry an unwanted load.

Several other spider crabs that warrant mention are the sheep crab, *Loxorhynchus crispatus,* the kelp crab, *Epialtus productus,* the long-armed spider crab, and the dwarf crab, *Pelia tumida.* The sheep crab occurs along the coast of southern California. The carapace is thick and tapers to a long prominent rostrum, the legs are long, the chelipeds stretching fully two feet, and the body is covered with long tubercles and spines and with short, bristly hairs. The kelp crab is found along the coasts of California and Oregon, where it lives among the seaweeds on rocks just below low-water mark. Olive green in color to simulate the kelp among which it lives, it has a smooth, square-shaped carapace with two spines on each side and a prominent toothed rostrum. In the long-armed spider crab, the carapace is longer than it is broad with three elevations all covered with spines. The surface is pitted and granulated, the rostrum is pointed downward, and the chelipeds are very long, the margins armed with spines. The long-armed spider crab lives among the rocks from Cape Cod to Florida. The dwarf crab is a small spider crab, measuring only about three-quarters of an inch long and about one-half inch wide. The carapace is pear-shaped, the rostrum is divided into two branches, and the legs are fringed with stiff hairs. It occurs from Santa Monica Bay, California, to Mexico.

The giant Japanese crab *(Macrocheira kaempferi)* is not only the largest of the spider crabs but is the largest of the entire group of arthropods. When its pinching claws are outstretched they measure eleven feet from tip to tip. It is found in the waters around Japan.

True Crabs–Commensal Crabs The Dutch naturalist Van Beneden was the first to use the word *commensal* (1876) when he referred to animals that share the same food *("Le commensal est simplement un compagnon de table"),* but today the word has come to have a broader meaning and includes all intimate associations between different species that are not parasites or symbiotes. In other words an association of two different species living more or less together without injury to either. Thus there are crabs that live in the mantle cavity of living bivalve mollusks or in wormholes of living marine worms without harming them except perhaps to deprive them of some of their food. The commensal crabs (family Pinnotheridae) are small in size, with a usually nearly circular carapace and with small eyes that are often rudimentary.

One species of these crabs is the oyster crab *(Pinnotheres ostreum)* (Fig. 37), so named because it lives in partnership with the oyster, that is, the female does, the male being free-swimming. The female has a thin, whitish, transparent carapace; the male has a firmer shell, is dark brown above, with a dorsal stripe and two conspicuous spots. It is smaller than the female. The oyster crab is particularly abundant in oysters from the Chesapeake Bay area. Another species is the mussel crab *(P. maculatus).* It lives with the edible mussel, scallops, and other bivalve mollusks and is common from Cape Cod southward.

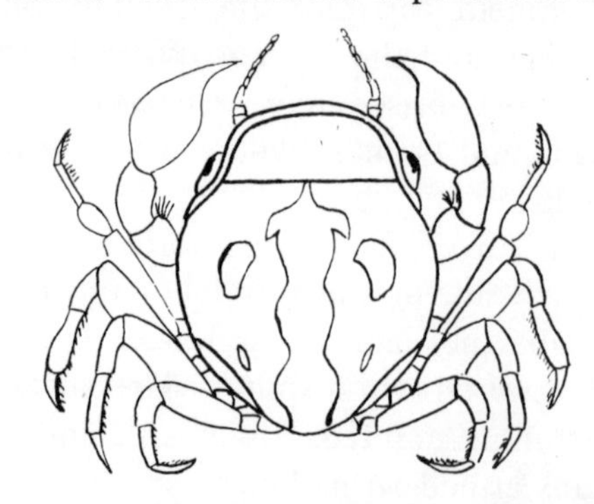

Figure 37. Oyster Crab, Male, *(Pinnotheres ostreum)*

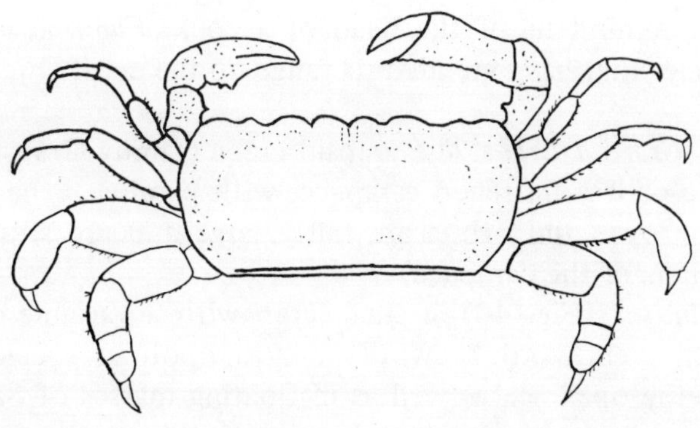

Figure 38. Parchment Worm Crab, Female *(Pinnixia chaetopterana)*

A third species, the parchment worm crab *(Pinnixa chaetopterana)* lives in the tube of the parchment worm, so named from the parchmentlike character of its tube. Both the carapace and legs of this species are very hairy. It is found from Cape Cod southward (Fig. 38).

Both the male and female crabs live in the tubes of the worm, which are found embedded in sandy mud with two chimneys extending an inch or more above the sea bottom, and are seldom found outside the tubes. As a matter of fact, as the crabs increase in size, they would find it impossible to escape from the tubes, the chimney-openings being too small to permit their egress. The crabs feed on particles of small and microscopic organisms that are pumped through the tube by the rhythmical contractions of the worm. A related species *(P. cylindrica)*, found from Chesapeake Bay southwards, is a commensal in the tubes of the lugworm. Several members of the same genus found on the Pacific coast are commensals with various bivalves and marine worms.

True Crabs–Box Crabs In the box crabs (family Calappidae) the cephalothorax is rounded and is covered with tubercles and granulations. The antennae are quite small and the abdomen is completely hidden under the thorax. The box crabs are widely distributed throughout the world in tropical and warm temperate seas where they conceal themselves in the crevices of rocks or by burrowing in the sand.

The box crab, *Calappa flamma,* which lives on sandy and muddy bottoms offshore from North Carolina southward, has a carapace that is broad and straight on the posterior side and curved on the anterior side, narrowing to the front. The posterior side has many denticulations. The pinching claws are large, broad, and flattened with a toothed crest on the upper border and are adapted to fit compactly against the front of the body so as to protect it completely. Indeed, when they are folded and the small legs are withdrawn beneath the carapace, the crab is shut up as if in a box and then resembles a shell. The crested claws, incidentally, remind us of the head of a cock. *Flamma* is actually a tropical species that has extended its range up the coast (Fig. 39).

True Crabs–Grapsid Crabs The grapsid crabs (family Grapsidae) have a square, more or less flattened carapace, with parallel or nearly parallel margins. The eyes and orbits are fairly large and are situated at the anterior corners of the carapace.

Planes minutus (Fig. 40), a small crab with a variable color, being mottled from yellow to brown, is a cosmopolitan species, living in seaweeds in the open sea as well as in floating masses of Sargassum. It often mimics the masses of seaweeds in which it lives, thus making itself invisible.

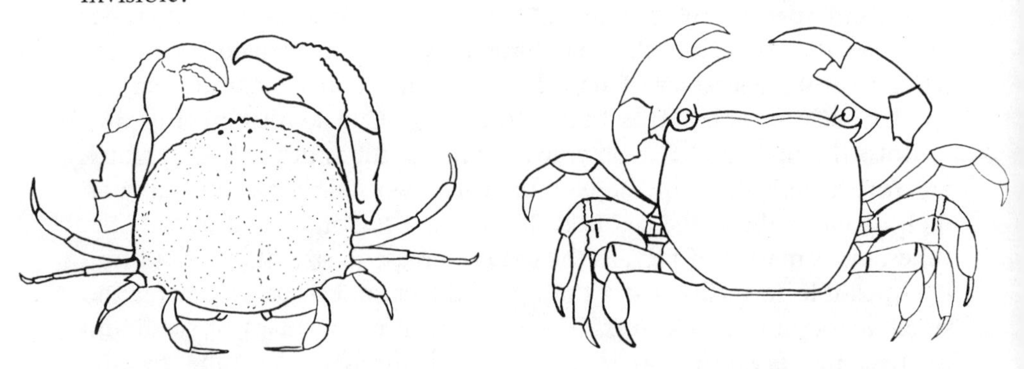

Figure 39. Box Crab *(Calappa flammea)*

Figure 40. *Planes minutus*

The striped rock crab *(Pachygrapsus crassipes)*, so named because it is striped with red, green, and purple and found on rocks as well as on sand, is very common on the Pacific coast from Puget Sound to San Diego. Two other species found on the Pacific coast are the purple shore crab and the yellow shore crab *(Hemigrapsus nudus* and *H. oregonensus)*. Both range from Alaska to San Diego and both are quite common. They live in sloughs of salt or brackish water. The yellow shore crab has a nearly square carapace, the anterior half of the side margins with two rather deep indentations, making two spinelike projections that bend forward. The four posterior pairs of legs are more or less hairy and the general color of the crab is yellow. The purple shore crab is purplish or brownish in color, the claws with red spots. Its walking legs lack hairs and it is a little larger than the yellow shore crab.

Another grapsid crab is the wharf crab, which is found on wharf piles and under logs and driftwood along the Atlantic coast from Chesapeake Bay southward.

True Crabs–Land Crabs The land crabs (family Geocarcinidae) are species of crabs that have become adapted to live on land where they inhabit burrows often far from open water of any kind. In addition to the series of gills present in all crabs, the land crabs have the portion of the shell covering the gills considerably inflated and lined with a thick membrane richly supplied with blood vessels that enable them to make use of atmospheric oxygen in much the same manner as do the lungs of mammals. The land crabs, as befits their name, spend most of their time on land, but they have to make periodic visits to the sea to breed and to permit the larval crabs to develop. Land crabs live essentially in tropical and subtropical countries, where they sometimes inflict serious damage to crops during their nocturnal wanderings. The white land crab of the American tropics often attains a size of nearly five inches across the back, with its claws as much as six inches long.

True Crabs–Sponge Crabs The sponge crabs (family Dromiidae) are the most primitive of true crabs. Their abdomen is not quite so reduced as in the other true crabs, their carapace is longer than wide, their eyes and antennules can be drawn into orbits, their last two pairs of legs are reduced in size, and their sixth pair of pleopods are lacking or vestigial. These crabs are called sponge crabs because they have the habit of covering themselves with sponges or pieces of sponges or holding them over their heads or backs for concealment. The sponge crab, *Dromia erythropus,* which may be found along the southern part of the Atlantic coast and is typically a West Indian crab, not only cuts out a piece of sponge that it holds over its head, but to make such concealment more effective it usually settles down in the cavity made by cutting out the sponge fragment, and in such a manner that the edges of the fragment fit into the cavity from which it was taken and thus is continuous with the rest of the sponge.

A curious little crab of the same family found on the coast of Florida *(Hippoconcha arcuata)* carries half of a bivalve shell over its back instead of a sponge. It makes use of its fifth pair of thoracic legs, which are bent over the back, together with the fourth pair and the spiny edge of the carapace, to hold the shell in position.

Several other species of true crabs that we might mention in bringing our account of the true crabs to a close is the calico crab or "Dolly Varden" *(Hepatus epheliticus)* (family Matutudae), a strongly colored tropical species ranging north to North Carolina with conspicuous rounded or irregular dark-colored spots on the back; the masked crab *(Corystes cassivelaunus),* a European species with markings on the carapace that somewhat resemble a human face; the purple crab of California *(Randallia ornata)* (family Leucosiidae), with a hemispherical carapace and mottled with blotches of red; and the South European river crab *(Thelphusa fluvialitis)* (family Thelphusidae), which is a good example of the freshwater crabs that are abundant in most of the warmer regions of the world.

Various species of crabs are an important source of food for man. Most important and valuable are the edible crab of British and European coasts *(Cancer pagurus),* the blue crab *(Callinectes sapidus)* of the Atlantic coast of the United States, and the Dungeness crab *(Cancer magister)* of the Pacific coast of the United States, the edible crab (Cancer) fishing of Europe ranking next in importance to the lobster industry. In India the crustacean fisheries outweigh in quantity and value the yield of all other fisheries in that country, the crabs being second in importance to the shrimps. And the swimming crabs, Scylla and Portunus, are among the most important sources of seafood throughout the entire Indo-Pacific region. Crabs are usually caught in baited traps or "pots," but the blue crab of the American Atlantic coast is often taken on baited lines.

In the order Decapoda there are a number of crustaceans that are not lobsters, shrimps, or crayfishes or true crabs either, though many are called crabs. They are intermediate between the lobsters, shrimps, and crayfishes on the one hand and the true crabs on the other and are often referred to as anomalous forms. They include the porcelain crabs, hermit crabs, and sand bugs.

In these forms the abdomen is more or less reduced and is usually bent forward beneath the thorax; it is horizontally extended only in rare forms. The abdomen is not too well armored, the lateral plates being either small or absent altogether. The third pair of legs are unlike the first and always without claws, and the last pair are unlike the third and either being turned upward on the dorsal surface of the animal or carried within the gill chamber. The gills are usually few in number and uropods are usually present but often reduced in size.

Porcelain Crabs The porcelain crabs (family Porcellanidae) often have a porcelainlike carapace, hence their name. Their bodies are small, rather flattened, and their abdomens are folded close against the thorax. The

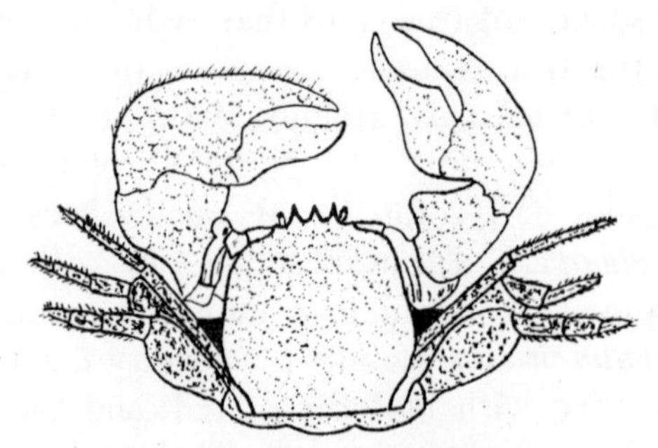

Figure 41. ***Porcellana sayana***

pinching claws are large, broad, and flattened, the first three pairs of walking legs are well developed, but the last pair are very small and are directed forward alongside the carapace. They are generally tropical in distribution, but several species are found along the southern Atlantic coast such as *Porcellana sayana* (Fig. 41). This species occurs from South Carolina southward. The carapace is a little longer than broad, with three acute, toothlike projections between the eyes, the middle one the largest, and two on each anterior side. The chelae are fringed with hairs on the edges and the walking legs are somewhat hairy. In color it is reddish or rusty brown with white spots or lines. It is found among oyster shells and often in shells occupied by hermit crabs.

A second species, *Porcellana soriata,* with a somewhat hexangular carapace, is often found in the canals of sponges washed on southern shores, and a third species, *Polyonyx macrocheles,* with a transversely oval carapace, and chelipeds unequal in size and quite long, ranges from North Carolina to Rhode Island and southern Massachusetts where, however, it is rare.

Hermit Crabs Everyone, or so it seems, has heard or read of the hermit crabs, whose posterior part of the body or abdomen is not protected by a crustaceous covering and which therefore have to seek protection for their soft and defenseless abdomens by inserting them into some hollow object, which is usually the shell of a gasteropod mollusk, as the whelk or periwinkle, though other objects are also used, such as plant stems and sponges. These animals appear well fitted to adjust to their surroundings, the body becoming modified according to the protective covering that they utilize; thus in those that live in shells the abdomen becomes spiral to conform to the convolutions of the shell.

In the hermit crabs the abdomen is usually not symmetrical but elongate, tapering, and soft. The antennae and eyestalks are long, and the first pair of thoracic legs are much larger than the others and are provided with claws. The right one is usually much larger than the left and, in addition to the usual functions of capturing and crushing prey, serves as a lid to close the opening of the shell when the crab retires completely within it. The second and third pairs of legs end in simple hooks and are used for walking and dragging the crab along when it moves over the ground, the fourth and fifth pairs are small, and the abdominal appendages are more or less atrophied, the sixth or last pair being modified to hold the animal in the shell that it occupies. In the female some of the abdominal appendages are adapted to carry the eggs.

Every now and then the hermit crab has to go house hunting, to find a larger shell to accommodate an increase in size as it grows. Many amusing stories have been told of the troubles that the animal often experiences in its search for a new home. Sometimes it finds a shell that appears suitable on first inspection but more often it has to try several before it finds one to its liking. House hunting is a careful and meticulous task, and a potential shell is carefully inspected to be sure that it is free from debris. Satisfied that it is clean, the hermit then wriggles into it to determine whether its abdomen fits the whorls of the shell. If it doesn't then it has to go in search of another, and it may try half a dozen or more before it finds a suitable one. The crab inspects a potential shell with its antennae and legs, and on deciding to occupy it, withdraws its abdomen from the old one and enters the new one so quickly that it is difficult to follow the crab in changing its home. If, after walking around with its new dwelling for a while, should it prove unsatisfactory in any way, the crab goes in search of another, but if satisfied, it then settles down in its new home and never ventures out of it until further growth makes it necessary to again change its domicile or should some stronger crab dispossess it. Sometimes the shells of hermit crabs become the home of hydroids and sea anemones that live with the hermits as commensals.

A common hermit crab, found from Maine to Florida, is *Pagurus pollicaris.* It is a rather large hermit, pale red in color, with the chelae covered with tubercles. It occurs at low-water mark on rocky and shelly bottoms of bays and sounds and lives in the shells of the common whelk (Busycon) the sand-collar snail (Natica), the moon shell (Polinices), and those of the large marine snails.

Another rather large species is *Pagurus bernhardus* (Fig. 42). It is bright red in color and rough and hairy. It ranges from Cape Cod northward and is also circumboreal. It lives in the shells of the giant

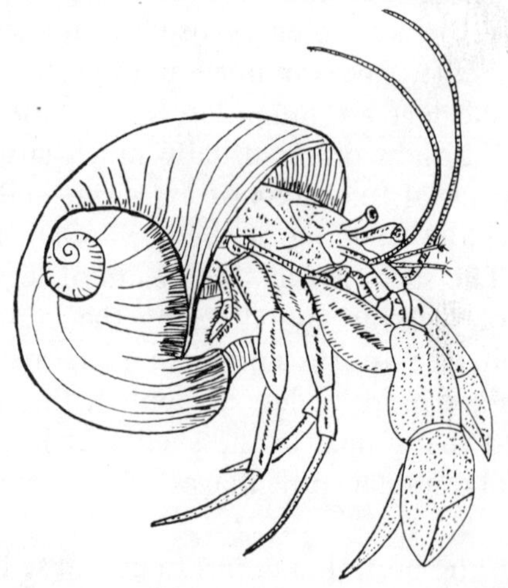

Figure 42. Hermit Crab *(Pagurus bernhardus)*

whelk (Fulgur) and the moon shell (Polinices). A quick-moving little hermit is *Pagurus longicarpus.* It has long chelipeds and its eyes are dilated at the end of the eyestalks. This species can easily be recognized by its very light color and metallic luster. It is an abundant crab from Maine south to South Carolina and may be found in rock pools and behind sand bars and also on sandy and muddy bottoms, where the water is shallow and sheltered, living in the shells of periwinkle (Littorina) and mud snails (Alectrion).

The hermit crabs are distributed throughout the world, and though mainly marine, some species of tropical regions are land forms. On our own Pacific coast we have some very large hermits, twelve to eighteen inches long.

Robber Crabs One of the most unusual of the terrestrial decapod crustaceans is the robber or coconut crab *(Birgus latro)* of the Indo-Pacific Islands. It is allied to the hermit crabs but has dispensed with a portable dwelling because the upper surface of its abdomen has become covered with calcified plates that provide adequate protection against injury from above. Adult robber crabs may attain a length of eighteen inches and can be quite dangerous if approached without caution. Their large claws can easily amputate a finger, and their sharply pointed walking legs are also effective weapons.

The robber crabs live in holes in the ground and rarely go into the water. They breathe atmospheric oxygen, their gill chambers having been converted into lungs by the presence of a network of blood capillaries. Their food consists of coconuts, which they tear apart with their heavy claws. They are popularly supposed to climb the trees to get at the fruit, but the matter of their tree-climbing ability appears to be in dispute.

Sand Bugs Some peculiar-looking crabs having a somewhat oval elongate body with a very short abdomen and long, featherlike antennae belong to the family Hippidae, a common species of the group found along the Atlantic coast being *Hippa talpoida,* popularly known as the sand bug. These crabs have a cylindrical carapace, simple first legs, second to fourth pairs of legs with the last segment curved and flattened and thus modified for digging in the sand, and a small rostrum or one absent altogether.

The sand bug (Fig. 43) is most uncrablike in appearance. The carapace is about one and a half inches long, convex, yellowish white, and nearly smooth. The abdomen is long and pressed under the body, the eyes, set on the ends of long, slender stalks, are small, and the antennae are plumelike.

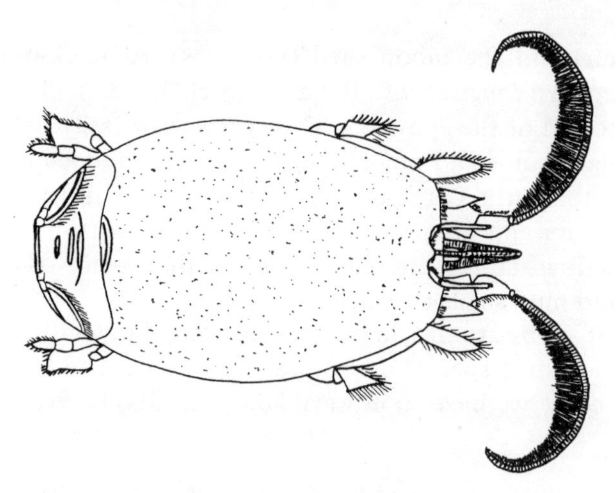

Figure 43. Sand Bug or Sand Crab *(Hippa talpoida)*

The sand bug lives on sandy beaches at or near low-water mark where it is exposed to the action of the waves. It is capable of digging quickly into the loose and shifting sands, using the short and stout second, third, and fourth thoracic legs and the appendages of the sixth abdominal segment for the purpose. When the appendages are folded under the carapace the little animal looks like an egg, its body being ovate, about half as broad as long, and with the sides forming a somewhat regular curve.

The sand bug is gregarious, and we usually find it in great numbers, but they are not captured readily, since they are able to disappear rapidly in the sand. Sometimes they may be seen swimming in the tide pools. As the waves come in and cover their burrows, they emerge and quickly scramble to a higher level, where they bury themselves anew, only to emerge again as the larger waves again cover their burrows and to scramble to still higher ground. When they are abundant or where they occur in large colonies, the edges of the waves appear to be alive with them. They feed on organic particles found in the sand, which they swallow as such, their mouths not being adapted for chewing. There are several species, one, *Hippa analoga,* which is similar to *talpoida,* but broader and flatter, being common along the Pacific coast from Oregon to Mexico.

Stone Crabs The stone crabs are free-living, crablike animals (family Lithodidae) having a globular and spiny body with the abdomen bent under the thorax and unsymmetrically composed of heavy calcified plates, the fifth pairs of legs reduced in size and folded in the gill chamber, and with a slender and pointed rostrum having thornlike spines on its sides. They live in cold water toward the poles or in the deep sea. A representative species is *Lithodes maia,* whose principal habitat is the coasts of the North Pacific, but it also occurs in the North Sea and on the fishing banks off the coast of Maine, where it is often taken in lobster pots. It is yellowish-red in color, lighter underneath.

Shrimps and Prawns Most likely what most of us know about shrimps is what we have learned from eating them in a restaurant; few of us have probably ever seen a live one. In the shrimps the body is compressed or more or less cylindrical and elongate with a fully developed abdomen. At the end of the abdomen there is usually a swimming fin formed by the uropods and telson. The rostrum is long, often longer than the thorax, and the eye-stalks, antennae, and legs often attain an extraordinary length. The first antennae have two or more flagella and the second have usually a large antennal scale. The scales, together with the long bases of the first antennae or antennules and the prominent eye-stalks, combine to make the head a broad and conspicuous feature. In some species the first three pairs of legs are provided with claws. In some species the young are hatched as nauplii but in most forms they appear in an advanced larval stage. There isn't a too well-established difference between shrimps and prawns; generally speaking very large shrimps are known as prawns.

Unlike the crabs and lobsters, which are more or less solitary creatures, shrimps are gregarious by nature and swim and congregate on the ocean floor in large schools. Inasmuch as they lack a heavy protective armor and powerful claws, they are less inclined to fight than their better-equipped relatives and when faced with danger "take to their heels" rather than stand and give combat to an enemy. They swim forward by using their abdominal appendages or dart backward by means of their tail fins in the manner of the lobsters. The shrimps feed on small crustaceans, worms, fish, and the like and in size range from that of a mosquito to about twelve inches in length.

In the spring and summer the shrimps move from the shallow coastal waters to deeper depths to breed and lay their eggs. The females lay from half a million to a million eggs, and unlike the lobsters and crabs that carry them on their abdominal appendages, they release them directly into the water. The eggs, which are small and round, barely visible, hatch in less than a day on or near the bottom into young that look like tiny mites. These swim or drift with the currents for many weeks as meanwhile they change in form, and eventually they reach shallow waters near the shore where they stay until they have reached maturity when they move out to sea again. The tail or abdomen is the part we eat. A pound of tails contains four to five hundred calories and is fairly rich in vitamins A and D.

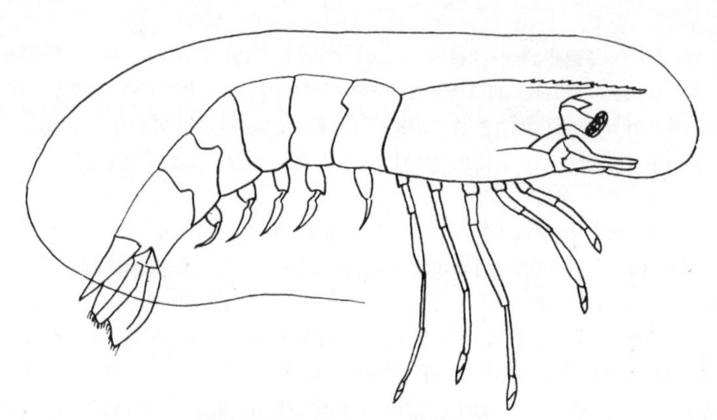

Figure 44. ***Peneus setiferus***

The most important market shrimp is the species *Peneus setiferus* (Fig. 44) (family Peneidae). When fully grown it measures about six inches long. There is a ridge or crest along the center of the carapace; it ends in a long, pointed, toothed rostrum. The antennae are a foot or more long; the first three pairs of legs have claws; and the abdominal appendages, as well as the lateral margins of the abdominal segments, are fringed with hairs. This shrimp is found from Virginia southward.

Generally associated with *P. setiferus* is *P. braziliensis.* It is not so abundant and less commercially valuable than *setiferus,* though it forms a part of the shrimp supply in the market. It differs from *setiferus* in having the crest on the carapace extending the entire length of the carapace. *Braziliensis* has a more northward range, being found as far north as Cape Cod, where it is often seen in brackish water or even where the water is quite fresh.

Members of the family Pandalidae are rather long-bodied shrimps with a long, slender, upturned rostrum and very small claws on the second pair of legs. *Pandalus borealis* is a circumpolar species and is found from Massachusetts to Greenland and from the Behring Sea to the Columbia River. Its rostrum is one and three-fourths times as long as the carapace and is armed with sixteen to twenty-one teeth.

Shrimps of the family Hippolytidae have claws on the first two pairs of legs, well-developed eyes, deeply cleft mandibles, and the second maxillipeds with a very short seventh segment. *Hippolyte pusiola,* which is often common on rocky bottoms in shallow water from Vineyard Sound northward and also in Europe, is pale gray or white, brightly spotted with red.

Included in the family Palaemonidae are shrimps with somewhat small claws on the first two pairs of legs, the second pair usually being the larger. The rostrum is long and compressed and armed with teeth. The species, *Palaemonetes vulgaris* (Fig. 45), is generally known as the

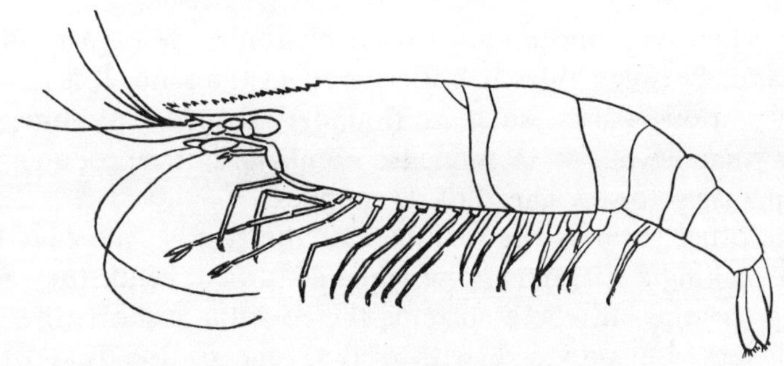

Figure 45. Common Prawn ***(Palaemonetus vulgaris)***

common prawn. It averages about one inch and a quarter in length, has a translucent body with brownish spots, and is found among eel-grass in brackish water, and also in pools and ditches on muddy shores from Massachusetts to Florida.

A member of the family Cragonidae and common along both the Atlantic and Pacific coasts is the species popularly called the sand shrimp. The shrimps in this family usually have a short, somewhat dorsally flattened rostrum and claws on the first two pairs of legs. The sand shrimp, *Crago septemspinosus,* is translucent, pale gray, with small, star-shaped spots, closely resembling the sand. It is broad at the anterior end and tapers to a sharp point at the posterior end. A pair of broad, divided appendages on the next to the last abdominal segment together with the telson form a fanlike swimming tail. As in all shrimps the antennae are long and have platelike scales at the base, which are fringed with hairs (Fig. 46).

The sand shrimp is abundant on sandy shores at low-water mark and in shallow water below tide mark, as well as among rocks and seaweeds. Its gray color imitates that of the sand so well that it is practically

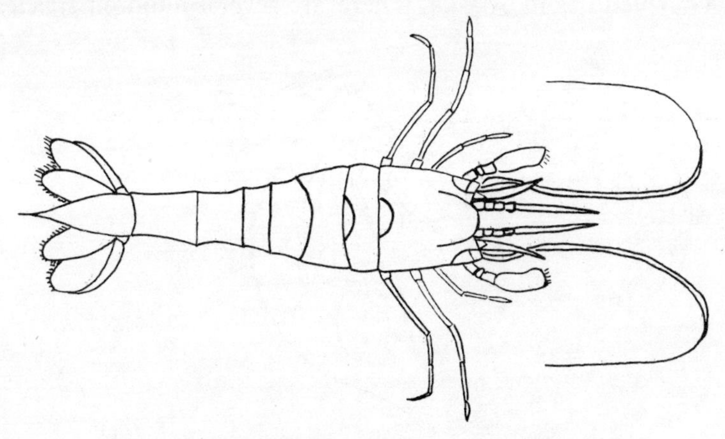

Figure 46. Sand Shrimp ***(Crago septemspinosus)***

invisible when lying motionless on the bottom or when partially buried in the sand. Between tides it buries itself in the sand. It is relished as a food by various fishes, such as flounder, striped bass, bluefish, and kingfish, that devour it in immense numbers, but reproducing prolifically it manages to stay ahead of its enemies.

Unlike other shrimps the members of the family Apheidae have the habit of making a snapping noise when disturbed, hence they are called snapping shrimps. In these shrimps the rostrum is short, the first two pairs of legs are provided with claws, one of the first pair being enormously large and stout, the claws of the second pair being small and of the same size, and the eyes are small and are usually covered by the carapace. The snapping sound is made by the joint of the large claw of the first pair of legs. The snapping shrimps either burrow in the sand or mud or are commensal with other animals, such as *Synalpheus lonicarpus,* which lives in the cavities in sponges. It ranges from North Carolina southward.

Most of the snapping shrimps are tropical, but another species, *Crangon heterochelis* (Fig. 47), occurs from Virginia to Florida and also on the coast of California. It is especially common in the oyster reefs off North Carolina. During World War II the snapping shrimps on the sea bottom off the coast of California were a nuisance to sonarmen and made submarine listening devices useless with their almost constant snapping noise, for whenever a small fish or some other prey ventures too near the burrow of one of these shrimps the shrimp fires a "shot" that stuns the victim, whereupon it is dragged into the burrow and devoured.

Shrimps occur in many parts of the world, and though they are usually thought of as warm-water shellfish, they are also found in northern seas. There are commercial shrimp fisheries in the waters of Norway, Greenland, and Alaska. There are several hundred species of

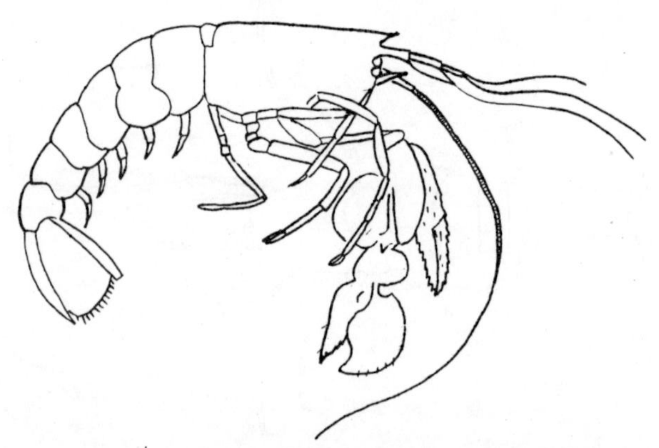

Figure 47. Snapping Shrimp *(Crangon heterochelis)*

shrimps, which, commercially, are the most important of the crustaceans as human food. The catch of species of Penaeus is believed to exceed in magnitude and importance that of all the other crustaceans taken for the market. The world's largest shrimp fishery is in the Gulf of Mexico, where several hundred million pounds are taken annually by fishermen of the United States, Mexico, and Cuba.

Shrimps are caught in several ways—with hand or cast nets, baited traps, haul seines, and with boat-drawn beam and otter trawls. The trawls, which are probably the chief method, consist of large, baglike nets that are dragged over the floor of the ocean, scooping up the shrimps in their path. Shrimps are marketed fresh, frozen, dried, canned, and cooked-and-boiled. Shrimp bran, made from dried heads and hulls, is sold for animal feed.

Crayfishes Anyone who has explored a pond or stream knows the crayfish (Fig. 48), for this animal is a familiar inhabitant of such places,

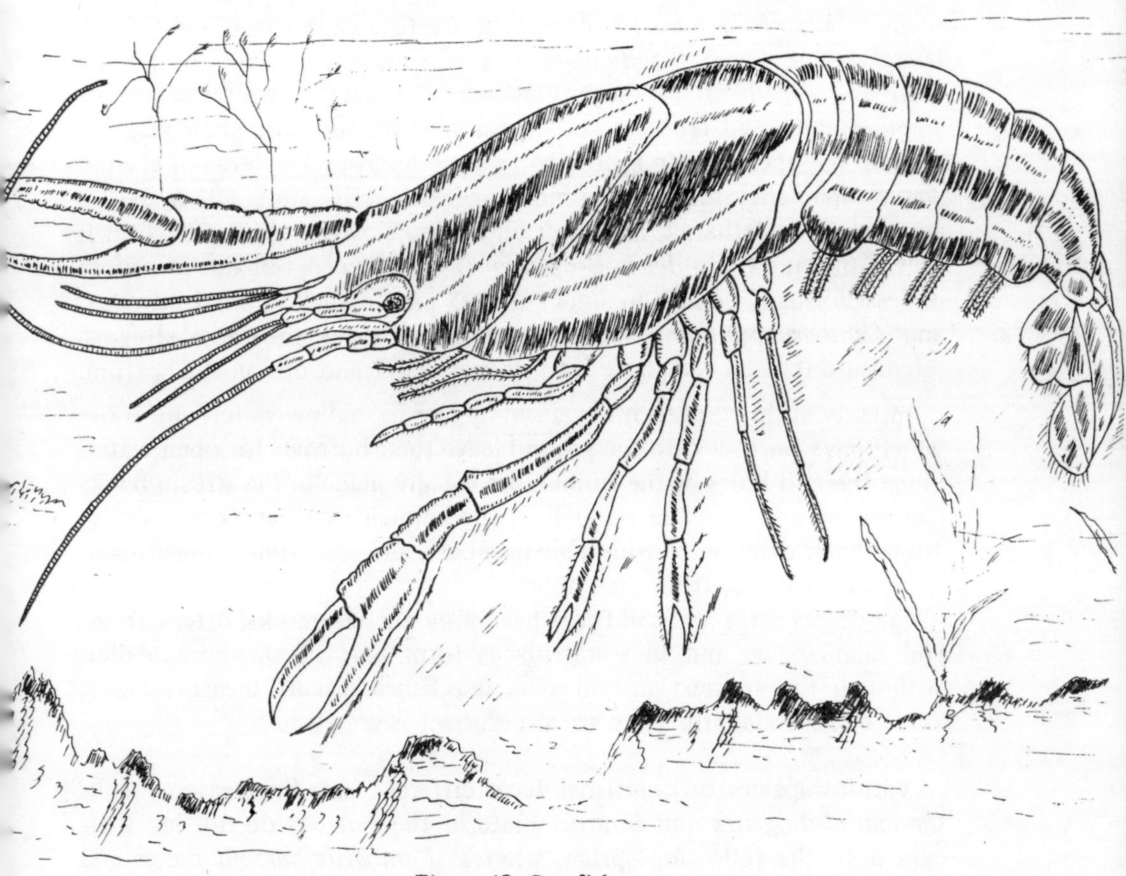

Figure 48. Crayfish

where it usually lives under a flattish stone or in a shallow burrow in the gravelly bottom. From such a sheltered situation it lies in wait for some passing fish, a water insect, or some other creature, which it clutches and tears to pieces with its large pincers, without exposing itself. The crayfish is also a scavenger, however, and feeds on all sorts of dead plant and animal matter.

Except for minor differences, the crayfish is so much like the lobster in structure and habits that it is often referred to as the freshwater edition of the lobster. Hence we shall dispense with a description of its structural appearance, since to do so would be merely repetitious. There are a number of species, and they are found all over the world. Some species, such as *Cambarus bartoni,* a common inhabitant of small, clear streams, and *Cambarus limosus,* which lives in rivers, do not burrow to any great extent, although all crayfishes dig into the mud and gravel bottoms as a means of securing protection from a number of dangers. These animals are more or less nocturnal, and in their native haunts they travel over the bottoms, hiding under stones or excavating shallow burrows in an attempt apparently to escape strong light. Such burrowers as *Cambarus diogenes* dig burrows that extend below the frost line and down to the ground water, and when winter sets in or when streams and lakes recede in summer they go into their tunnels, plug up the openings, and retire to their underground cisterns of ground water where they remain in a more or less inactive state. Other species are somewhat lethargic in winter and move slowly over the bottom if they move at all: *Cambarus propinquus,* for instance, remains in a more or less dormant condition under stones and in shallow mud burrows, and *Cambarus immunuis,* an inhabitant of lakes and slow streams, migrates out from the shore into deeper water and lies on the bottom.

In early spring crayfishes begin to appear in shallow water, and those which have burrowed in the ground leave their burrows for open water. Since the exit holes of the tunnels are usually made in the stream banks just below or above the water line, the animals may be seen emerging from them, often in considerable numbers, and sometimes appearing at the same time, one from each hole.

Crayfishes serve as food for fishes, birds, turtles, minks, otters, frogs, and salamanders and in some places form part of the human diet. Although the species vary in size, crayfishes usually measure about three inches in length. They are also known as crawfishes.

Various species of crayfishes have different breeding seasons. Thus *Cambarus diogenes* and *limosus* mate in the fall but do not lay their eggs until the following spring, whereas *Cambarus bartoni* mates and spawns the year round. The life history of *Cambarus diogenes,* which is somewhat as follows, is illustrative of that of other species. In this species the females lay their eggs within the burrows in April or later. Before laying her eggs the female cleans the underside of her abdomen with her pincers, after which she lies on her back, curves her abdomen upward, and then exudes first a gluelike secretion and then several hundred eggs out over the surface that stick to the swimmerets in a solid apron. The eggs require about a week to hatch, and during this time the crayfish, like the lobster, is said to be "in berry." At hatching the young crayfishes are not freed from the egg shells but remain attached to them by a filament connected with the telson. After one or two molts the young crayfishes take hold of the parental swimmerets by their own pincers.

The young crayfishes grow rapidly and live in the burrows with their parents until about the first of July. They are then a little less than an inch long and are ready to excavate burrows of their own. They continue to grow throughout the summer and then mate in autumn even though they have not yet attained their full size. The sexes can easily be distinguished by the shape of the first pair of abdominal appendages, in the male being used to transfer the sperms (Figs. 49, 50)

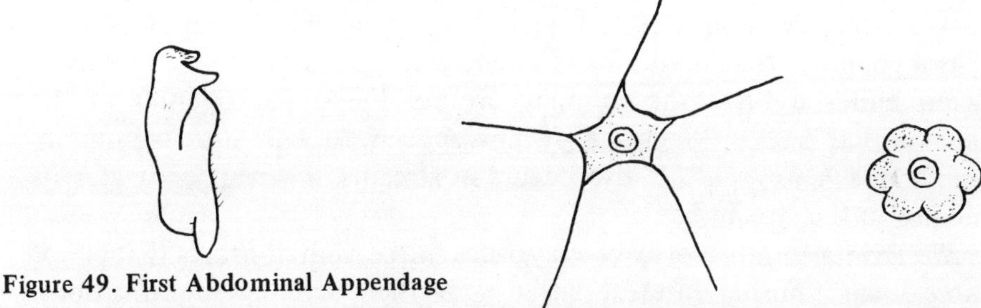

Figure 49. First Abdominal Appendage of Male *(Cambarus diogenes)*

Figure 50. Two Views of Sperms of Crayfish

to the female during the act of mating. In the summer the tips of these appendages are soft and flexible, and while they are in this condition the crayfishes are said to be in "second form" or juvenile, since the appendages first appear during this period. Toward the close of summer the young crayfishes of both sexes usually molt, the flexible abdominal appendages being replaced by stiff, horny ones. The crayfishes are now described as "first form" or adult. They are now ready for mating, which takes place in November. After they have mated, both sexes retire to their burrows, where they remain for the winter. During this time they grow very little or not at all. In the spring both males and females emerge from their tunnels, the females meanwhile having spawned and generally having been freed of their young. Sometime later, both sexes molt and in the males the stiff sexual appendages are replaced by the flexible juvenile ones like those of the previous summer, the sexual appendages, in other words, undergoing a seasonal change.

Crayfishes are able to form habits and modify them as has been shown by certain simple experiments. The chief factors in the formation of habits appear to be the senses of smell, taste, touch, and sight. They learn by experience and modify their behavior quickly or slowly depending on how familiar they are with given conditions.

The common crayfish of the Eastern states, occurring in small, clear, quiet streams, though occasionally in muddy ones, from Tennessee and the Carolinas to Maine, is *Cambarus bartoni* (family Astacidae). It can easily be distinguished by the smooth carapace of both sexes and the hooks on the first pair of abdominal segments of the male. In *Cambarus limosus* the carapace is spiny in both sexes and the body is hairy all over. It is common in the larger rivers and in ponds of eastern North America.

The species *Cambarus diogenes* is what is known as a chimney crayfish. It is widely distributed and is quite common throughout the United States east of the Rockies. It is found in swamps and meadows, often far from a stream, and is preeminently a burrowing species. It makes a burrow from one to three feet deep, at the bottom of which it digs a chamber that it fills with water, which serves as a safe retreat. It often builds a turret or chimney around the upper opening of the burrow that is several inches high. This species, though more commonly found in swampy places, also occurs in streams, where it excavates its burrow in the stream banks.

We have a number of cave crayfishes in the United States that are of more than unusual interest because of their striking modifications. They are all blind, their eyes being atrophied and their eyestalks more or less undeveloped, and they are all light colored, pigmentation being absent. Their claws are not too well developed but their antennae are long and highly specialized as tactile structures. For the most part they are small species. One of the cave crayfishes is *Cambarus pellucidus,* which is found in the caves of Indiana and Kentucky.

Crayfishes do well in aquaria, where they may be kept for long periods for study and observation or merely as pets. The aquaria should be supplied with a little mud and plenty of water plants. The crayfishes may be fed bits of meat, pieces of earthworm, and water insects, but care should be exercised in not overfeeding them.

Crayfishes are used rather extensively as an article of food in Europe, but in our country they have never become popular, though species in Louisiana and Oregon and in the Great Lakes area and in the Mississippi River drainage system are eaten to some extent. Since several species in the East serve as an intermediate host to the lung fluke *Paragonimus westermani,* which is parasitic in man and other carnivores, they should be thoroughly cooked before being eaten. In the South crayfishes are often a nuisance, since they denude young grain, sugar cane, and cotton seedlings and tunnel in dams and dikes. Their chimneys also are sometimes a threat to farm machinery, which they clog or render useless.

3
The Phyllopods

The phyllopods* are crustaceans with flat, leaflike thoracic appendages that serve the double purpose of respiration and swimming, the name phyllopod coming from two Latin words *phyllo,* leaf, and *poda,* foot. They have an elongated body with many distinctly marked segments and ten or more thoracic segments, a heart that extends through four or more thoracic segments, and a telson that usually bears a pair of elongate appendages called cercopods. The eggs may be produced by parthenogenesis or fertilization and hatch into a nauplius. The phyllopods are the fairy shrimps, the claw or clam shrimps, and the tadpole shrimps. There are over a hundred species. With the exception of the brine shrimp, they live in freshwater pools and usually swim on their backs.

Fairy Shrimps Visit any temporary spring pool, such as is formed by melting snow, and you may see the fairy shrimps (Fig. 51). They are rather unusual looking animals, measuring about an inch long, with a transparent body in which it is often possible to see the beating heart,

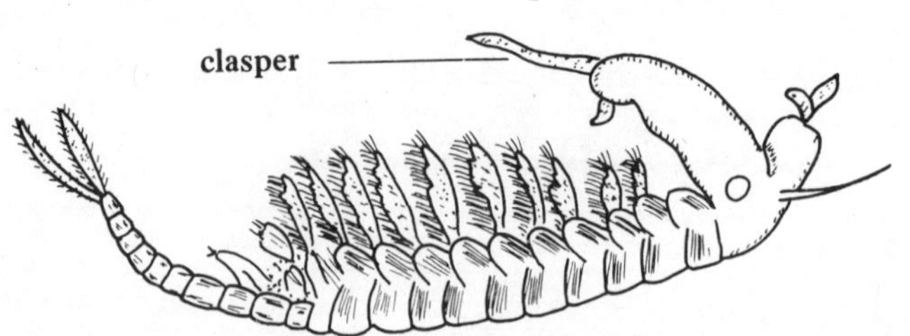

Figure 51. Fairy Shrimp *(Eubranchipus vernalis)*

*Formerly the phyllopods were grouped in the suborder Phyllopoda by some authorities or in the order Phyllopoda by others; more recently they have been grouped into three separate orders: Anostraca, Conchostraca, and Notostraca.

and are colored with all the tints of the rainbow, ranging from whitish to blue, green, orange, or red. The hind part of the body is slender and without appendages and is brightly colored by the blood, which is red with hemoglobin. Fairy shrimps usually swim on their backs and are remarkably graceful, moving through the water by beating their many leaflike appendages in a series of wavelike motions, then pausing to drift lazily about for a moment or so, only to dart forward again; and as they move through the water their waving plumes are plainly visible.

Fairy shrimps inhabit small, temporary bodies of water such as pools formed by melting snow or ice, potholes that are filled with water only in the early spring, or in ditches. They appear miraculously as it were overnight, grow to maturity, and reproduce within a few short weeks.

The fairy shrimps live on microscopic organisms, such as bacteria, protozoans, rotifers, diatoms, and algae, which is filtered out of the water and conveyed to the mouth by movements of their gill-feet or leaflike appendages. In turn they are eaten by amphibians, caddis worms, and the larvae of diving beetles. Since the water in which they live is only temporary, they have usually disappeared by the time other animals have become active and hungry and, except for those I have mentioned, have few enemies, although flamingoes are said to scoop up the brine shrimp, which, incidentally, were also harvested by the Indians.

Females, which can be recognized by the brood pouch on the side of the abdomen, appear to be more numerous than the males, and in some species no males are known, the young developing from eggs that have never been fertilized. In a typical New England spring the fairy shrimp population may reach its maximum within a week or two after the ice has left, and at that time females, carrying eggs in their brood pouches, and many mating pairs as well as maturing larvae may be seen swimming about. Mating fairy shrimps swim about together, the male holding the female to his ventral side with the peculiar claspers that are parts of his antennae modified as mating organs (Fig. 51), the most conspicuous features on the face of the male. Directly behind his gills, on the eleventh segment of the body, are tubelike appendages by which he transfers the sperm cells to the female.

Fairy shrimps can live only in cold water, and as the water warms or the pools dry up, they gradually fall to the bottom and die, but not before the females have deposited eggs, which lie dormant throughout the summer, or for a considerable period, in the bottom mud. Some of these eggs are thick-walled to carry the species through the hot, dry months and may be dried and frozen without injury. Some even seem to need drying before they can hatch. It has been reported that eggs kept in dried mud for as long as fourteen years developed and hatched after having been put in water.

The eggs eventually develop into a nauplius, which must undergo several molts before becoming adult. The eggs may hatch as early as January if there is water present, but usually they do not hatch until later.

The distribution of fairy shrimps is freakish and fraught with uncertainty. They may be numerous in one pool and entirely absent from one nearby, which to all appearances may be exactly like the first one. For several years they may appear as regularly as the seasons and then they may not be seen again for four or five years, although the conditions may be the same, or they may be abundant one season and then not reappear for several years. Although fairy shrimps are quite common their appearance or discovery is always an occasion for the naturalist or collector.

Unlike other phyllopods, fairy shrimps do not have a carapace. They have eleven pairs of swimming legs (seventeen to nineteen in two Arctic species), stalked eyes, and the second antennae with a single branch. A common but sporadic species is *Eubranchipus vernalis* (Fig. 51) (family Chirocephalidae), found throughout eastern North America and locally abundant in pools during late winter and early spring. It is semitransparent; stout and large, and about an inch long. The male can be recognized by his large frontal appendages or claspers, the female by her brood pouch.

The species *Streptocephalus seali* (family Streptocephalidae) is widely distributed throughout the United States, but the species known as the brine shrimp, *Artemia salina* (family Branchinectidae), which is confined to salt lakes and the brines of salt works, has a more limited range. Artemia (Fig. 52) measures about three-quarters of an inch in length, is semitransparent, and is red, pink, or greenish in color but is extremely variable, because of the concentration and the chemical nature of the salts in solution. Sometimes the water in which it lives is so dense that salt crystallizes on its body.

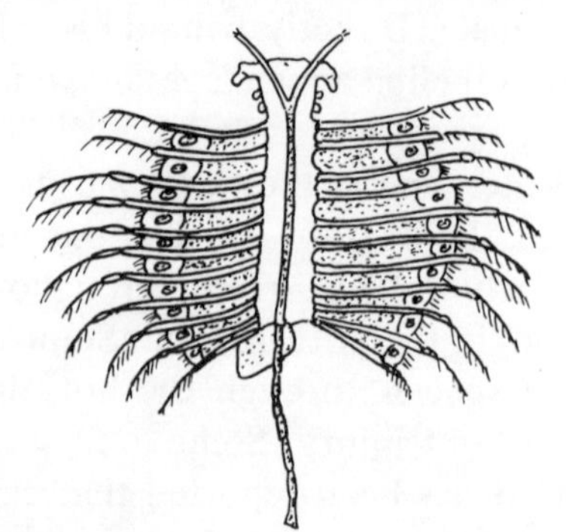

Figure 52. Brine Shrimp *(Artemia salina)*

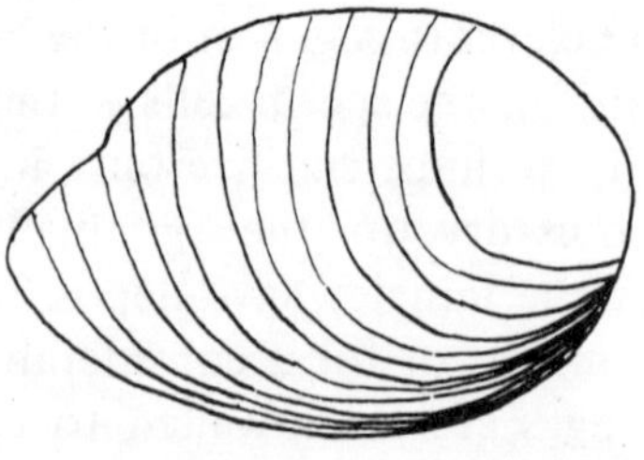

Figure 53. Clam Shrimp *(Limnadia lenticularis)*

Clam or Claw Shrimps Merely a glance at a clam shrimp and it is easy to see how it got its name, for its carapace forms two lateral, shelllike valves that completely enclose the body, making it look like a tiny clam. And to make the resemblance even greater, some species even have lines of growth like the true clams (Fig. 53). They have ten to twenty-eight pairs of swimming legs, stalkless eyes, and two-branched or biramous second antennae. Most clam shrimps live in the littoral zone of lakes, ponds, and temporary pools and prefer a somewhat warmer water than the fairy shrimps. Some species live in muddy water, a few in alkaline pools. Unlike the fairy shrimps, they swim right side up, that is, backside up, moving through the water with a rowing motion of their two long second antennae. Sometimes they move rather clumsily over the bottom, falling on their sides if their antennae stop waving or even burrow into it. They reproduce much like the fairy shrimps with the exception that the females carry their eggs in a mucous mass on the back, between the body and the shell.

Quite common in woodland pools throughout eastern and central United States is the species *Lynceus brachyurus* (family Lynceidae). It has a nearly spherical shell without any lines of growth, is about one-eighth of an inch long, and is pink in color with black eyes. The rostrum is very large and broad and the first pair of legs of the male are prehensile or grasping (Fig. 54).

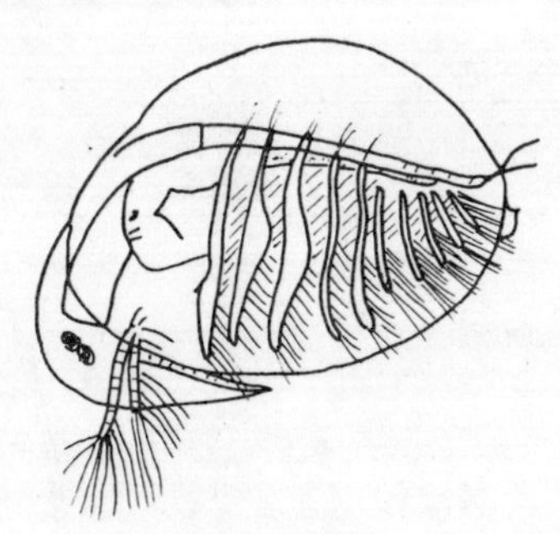

Figure 54. Clam Shrimp *(Lynceus brachyurus)*

Tadpole Shrimps The tadpole shrimps have an elongated body composed of many segments and an oval, low-arched carapace that covers the head and thorax. The eyes are sessile, that is, without stalks, the first antennae are short and threadlike with two or three flagella or threadlike projections, the second antennae are small or absent, and there are forty to sixty pairs of broad feet, and two long caudal cercopods. In appearance the tadpole shrimps are strangely like tiny horseshoe crabs of the seabeach (Fig. 55).

These crustaceans occur in freshwater pools and ponds where they swim backside up in a graceful, gliding manner. They also creep over the bottom and sometimes burrow into it. They feed on microscopic organisms as well as worms, mollusks, frogs' eggs, and dead tadpoles. The eggs, borne in egg capsules on the eleventh pair of appendages, hatch into nauplii.

Some species, such as *Lepidurus couesi* (family Apodidae), have the telson expanded to form a flat paddle, others, such as *Apus lucasaus* of the same family, do not. In the fomer the telson is keeled, in the latter it has three spines. Both species occur in the West. A species of the genus Triops (Fig. 56), of the West and the tropics, is sometimes a serious pest of rice fields in California, eating the leaves and stirring up silt, but in Mexico it is a valuable article of food.

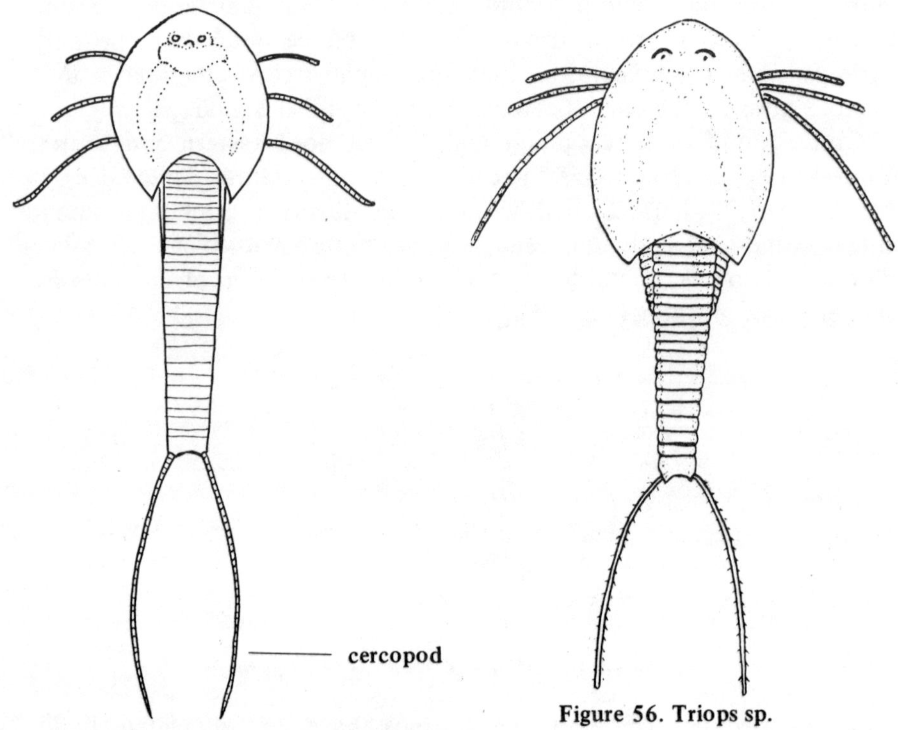

Figure 55. Tadpole Shrimp *(Apus lucasanus)*

Figure 56. Triops sp.

4
The Cladocerans

The cladocerans, otherwise generally known as water fleas, are minute crustaceans that occur in waters in amazing numbers. They are compressed from side to side and usually have a short and compact body that is generally covered by a transparent shell or carapace. The shell, which is a single folded piece, gapes ventrally, does not cover the head, and may be oval, round, elongate, or angular, sometimes with a posterior spine. The animals have four to six pairs of thoracic appendages or legs, the first pair of antennae is often minute, the second pair is very large with two prominent branches and is used for swimming, and the abdomen is small and is usually bent under the thorax. They have but a single median compound eye on a head that is often bent down and in profile appears birdlike; sometimes the head is provided with grotesque structures or extensions. The females have a large dorsal brood sac in which the eggs develop and that hatch into young that have the form of the adults. The winter eggs are often provided with an extra shell called the ephippium (Fig. 57), which

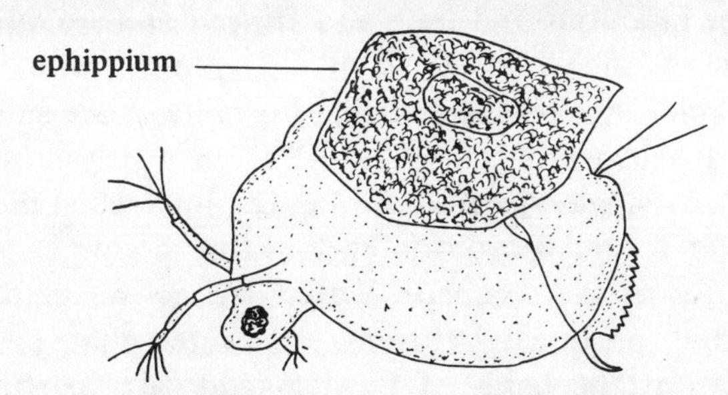

Figure 57. Ceriodaphnia with Ephippium

consists of two chitinous plates, much like watch crystals, whose edges fit together. There may be one, two, or more eggs in a single ephippium.

The smallest of the water fleas measures barely one hundredth of an inch, the largest is less than three-quarters of an inch. Despite their size they have highly organized systems for digestion, reproduction, and the like. Under a microscope they are beautifully transparent and you can observe every detail of structure without any particular equipment or preparation. A drop of water and a glass slide suffice. They are a delight to watch; you can see the beating of the large oval heart and the conspicuous eye jerking and rotating with the movements of the body. The species found in plankton are darker in color than the littoral species and are tinged yellowish brown, reddish brown, or even black. Some pond species have pink blood but most have pale yellowish blood.

The water fleas have long been favorite animals for study, as they may well be. In 1699, the Dutch naturalist Swammerdam described the common water flea as *Pulex aquaticus arborescens*—the water flea with the branching arms—and water flea it has been to this day.

Water fleas occur in all types of water with the exception of swiftly flowing streams and waters that have been polluted. A large cladoceran population is usually present in open water where it forms a part of the plankton society, though relatively few species belong to the plankton population of lakes and these usually have very transparent bodies. Some species inhabit calcium-poor waters; some live in acid bog waters; others, perhaps the greatest variety of species, occur in the weedy margins of ponds where they can be seen swarming near the shore and where thousands can be scooped up in a glass of water. A few species dwell on the bottom, where they scramble over the mud and are little given to swimming; others swim in half-open waters between floating leaves.

Whereas crabs and crayfishes use their second antennae as tactile organs or feelers, the cladocerans employ theirs for swimming. They are two-branched or biramous and long and fringed and are quite effective for the purpose. Some water fleas, like Daphnia, swim jerkily with uneven, short, quick strokes; others, like Bosmina, move smoothly with long strokes. Latonia will often take off from a resting position in a sudden leap with a single vigorous stroke. Daphnia when moving over a surface kick the forked posterior end of the abdomen (furca) out between the edges of the carapace. Water fleas move their leaf-shaped feet very rapidly, producing currents of water that pass over the respiratory valves at the bases of the legs and carry particles of food forward to their mouths.

Most water fleas feed on minute organisms such as protozoans, diatoms, and algae; some algae feeders provide a distinct service in holding in check the growth of algae in such places as reservoirs. Some of the larger cladocerans are primarily carnivorous, feeding on smaller water fleas, rotifers, and protozoans, which they capture with specially modified legs. In turn the cladocerans are eaten by hydras, insects, and wading birds, and are the staple food of fishes, their importance in this respect being due to their vast numbers and not their size. The water fleas have a tremendous reproductive capacity. Thus, for instance, one species, *Daphnia pulex,* produces a brood of eggs every two or three days and it has been estimated that in sixty days its descendants would number some thirteen billion individuals, that is, if they all lived.

The reproductive activities of the water fleas are fairly complicated and are greatly influenced by such environmental factors as temperature, food, and other conditions. We might use the seasonal history of the water flea Daphnia as typical. In early spring you will see only a few if any Daphnia in the ponds and pools, but as the water begins to warm up and the ice begins to break up, female daphnids hatch from the resting eggs that have lain dormant on the bottom all winter. These females soon produce thin-shelled eggs that may be observed in the transparent brood pouches. The eggs develop parthenogenetically into daphnids that are females like their mothers, males not appearing until later. The young remain in the brood sac for about two days and are then released. Before becoming fully mature they must molt several times, and by the time they are adults their own first clutch of eggs is ready in the brood chamber. Parthenogenetic eggs are produced throughout the spring and summer, in other words, one generation of females follows another every ten days or so, each animal bearing litters of young until the water is alive with daphnids.

The lack of food, due to the overcrowding by females, the approach of cold weather, and other unfavorable changes eventually bring an end to the procession of female generations. When this occurs the females produce eggs that develop into males. Mating then takes place and fertilized eggs are produced. The females produce only one or two or at the most very few of these eggs. They are relatively large, contain more yolk than eggs of the previous generations, and are enclosed in an ephippium. The ephippium is released, which, with its contained eggs, either drops to the bottom or floats upon the surface. The ephippia are highly resistant to drying and freezing and protect the eggs through the winter until the following spring when they hatch and the cycle is again repeated.

The thick-shelled eggs are often referred to as "winter eggs" and the thin-shelled parthenogenetic eggs as "summer eggs" because they frequently exist during these seasons. However, such terms are misleading, for species that normally produce resting eggs in the fall will sometimes under certain conditions also produce them in the summer, while there are species that regularly produce resting eggs in the summer. Certain water fleas are also called "summer species" or "winter species" according to the time when parthenogenetic females are plentiful. There are also perennial species in which sexual reproduction is reduced or absent. In such species resting eggs are not produced, the species being perpetuated year after year by parthenogenetic females.

The cladocerean population of a pond or lake increases and decreases in rhythms according to temperature and the season. The time it takes for a species to reach maturity after hatching also varies with the species; thus it has been found experimentally in the laboratory that *Moina macrocopa,* at a temperature of 68° F., will reach maturity in 114 hours whereas it will take *Daphnia pulex* 157 hours and *Daphnia longispina* 187 hours. Since its litter contained as many young as the other species, it produced a great number of progeny. Under a favorable temperature such a species could well become a dominant one. Further experiments also showed that at a temperature of 56° F. *Daphnia pulex* required 309 hours to mature but at a temperature of 77° F. it required only 100 hours. In all natural bodies of water, temperature constantly fluctuates and any given species of cladoceran must repeatedly encounter a temperature that either increases or decreases its number of generations, thus affecting its total population.

In any natural body of water, temperature changes not only with the season but also from day to day and for that matter from hour to hour. Even in the same pond, the temperature may change in different places according to the amount of exposure to sunshine or wind as well as other factors such as the amount of heat produced by decaying organisms. A sudden drop in temperature, at least low enough to make a species inactive, may result in the species dropping to the bottom to be smothered by the mud, thus wiping it out. Hence, whether a species thrives or not is largely a matter of temperature.

It is interesting to note that some species of cladocerans have very differently shaped bodies in summer and winter, a phenomenon known as cyclomorphosis. Most of such species occur in the open surface-water plankton where other organisms also go through similar summer and winter changes with usually a considerable variation in shape. For instance, in the females of such limnetic species as *Daphnia pulex* and *Daphnia longispina,* a population has a homogeneous "normal," or round-headed, form during the late fall, winter, and early spring, but, then, as the water becomes warmer and the population develops, there is commonly a progressive increase in the longitudinal axis produced by a general elongation of the head and the appearance of what is called a "helmet." By midsummer the helmets become fully developed and at that time may be quite bizarre. In late summer or early fall, the shape of the head begins progressively to revert to the "normal" head form, which is finally attained by late autumn. Cyclomorphosis may also involve changes in the size of the eye and in the length of the posterior spine.

It might be further added that the degree of summer helmet development differs widely in the same species, even in two neighboring lakes. In ponds and shallow lakes, cyclomorphosis is less pronounced and the degree of helmet development is relatively consistent from one individual to another, a condition that does not obtain in larger and deeper lakes where cyclomorphosis and the degree of helmet development are considerably more variable and where both strongly developed and poorly developed helmets, as well as intergrades, occur.

Various explanations have been offered to account for these seasonal changes in morphology but none seems to provide an adequate answer to what is rather a puzzling problem.

Of all the water fleas, the Daphnia (family Daphniidae) are undoubtedly the best known, more specifically the common Daphnia *(Daphnia pulex)*, which is widely distributed in America and in Europe. It is oval in shape, often reddish in color, with a prominent beak on the underside of the head, and is about a sixteenth of an inch long.

Although visible to the naked eye, it is only under the microscope that we can see what Daphnia really looks like. Look at it through the microscope if you can and you will find that it appears as shown in Figure 58. Talk about the privacy of goldfish, why, they haven't

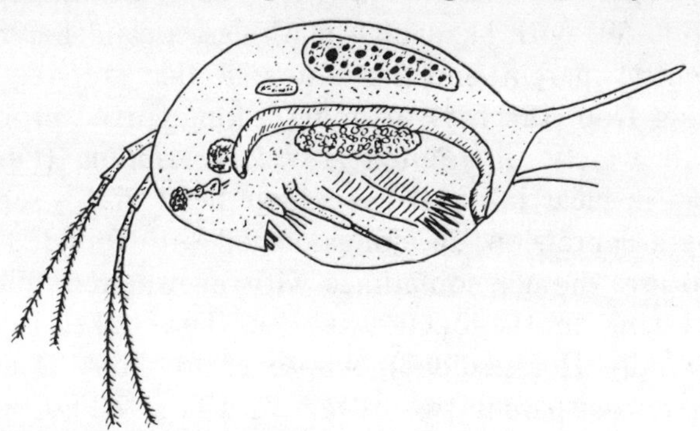

Figure 58. The Common Daphnia *(Daphnia pulex)*

anything to complain of compared to this little water flea, for Daphnia's life is just one big open book, since we can observe not only what it does outwardly but also what goes on inside—and this in spite of the shell that covers it like a knight in armor.

This little animal is a veritable dynamo, for if you watch it closely you will see that every part of it, both inside and outside, appears to be in motion. The rapidly beating legs are not legs at all as we understand them. They are not used to walk or swim with but to direct into the animal's mouth a stream of water containing microscopic animals on which it feeds and a certain amount of dissolved oxygen without which it could not live. Locomotion in Daphnia is effected, as I have already remarked, by the long, branched antennae, which are always in motion, giving to the animal a series of hopping movements through the water, for without such perpetual swimming the animal would drop to the bottom of the pond or pool in which it lives, since it is heavier than water and has no air bladder to keep it suspended in the water such as is the case with the fishes.

If you study the animal carefully you can see the passage of food through the alimentary canal, the blood flowing through the body, and even the contractions of the heart. And if you should happen to be viewing a female you will also see the eggs that she carries in her little pocket and that will appear as little dark specks massed together.

Living Daphnia is undoubtedly one of the best foods for tropical aquarium fishes. Although the animal is covered with a shell, the shell is soft and contains mineral elements that are very desirable, while the flesh itself is easily digested and is nutritious. Both live and dried Daphnia can be obtained at almost any pet store, a large species, *Daphnia magna,* being the one that is usually sold. Incidentally, there are some fifty species of Daphnia.

A very common water flea in the eastern and central United States as well as in Europe is *Bosmina longirostris* (family Bosminidae). It has an oval shell (Fig. 59) with hexagonal markings, a caudal spine projecting from the ventral margin, and first antennae that are greatly elongated and extending from the beak forming a long curved proboscis. Often common and widely distributed is *Sida crystallina* (family Sididae) which occurs in clear lakes. It has a large head that is separated from the body by a depression, an elongate shell with rounded ends, and is usually colorless though sometimes with brown, red, and blue spots (Fig. 60). Living in the open waters of large lakes is *Holopedium gibberum* (family Holopedidae), whose entire body is enclosed in a large globular, transparent two-valved shell (Fig. 61). Common in the eastern United States and in Europe, *Simocephalus vetulus* (family

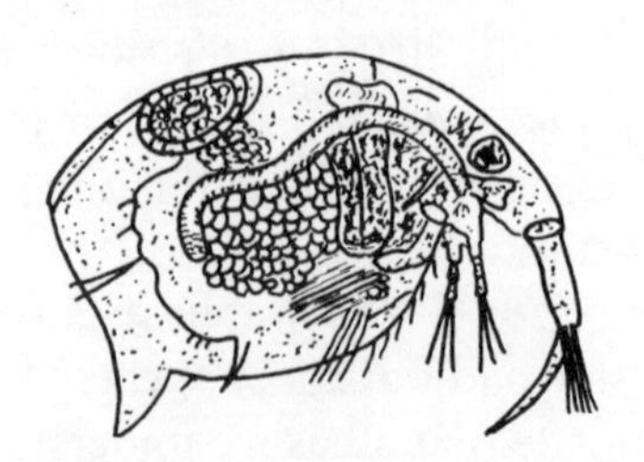

Figure 59. ***Bosmina longirostris***

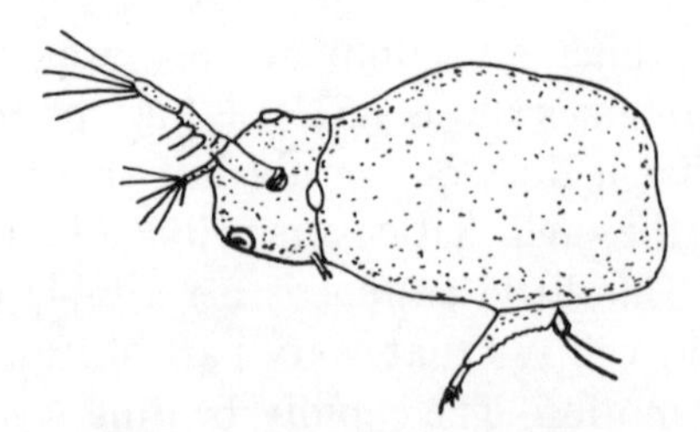

Figure 60. ***Sida crystallina***

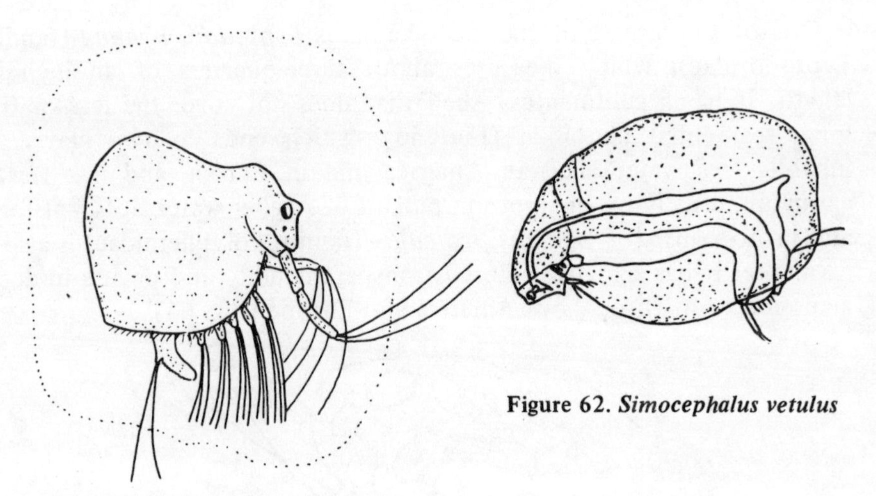

Figure 61. *Holopedium gibberum*

Figure 62. *Simocephalus vetulus*

Daphniidae) has a large, short, high body that is obliquely cut off behind (Fig. 62). Also with a body cut off at the hinder end is *Scapholebris mucronata* (family Daphniiae) but unlike Simocephalus, which does not have a caudal spine, Scapholebris has a pair of caudal spines that, however, are sometimes very short. Both Holopedium and Scapholebris swim upside down.

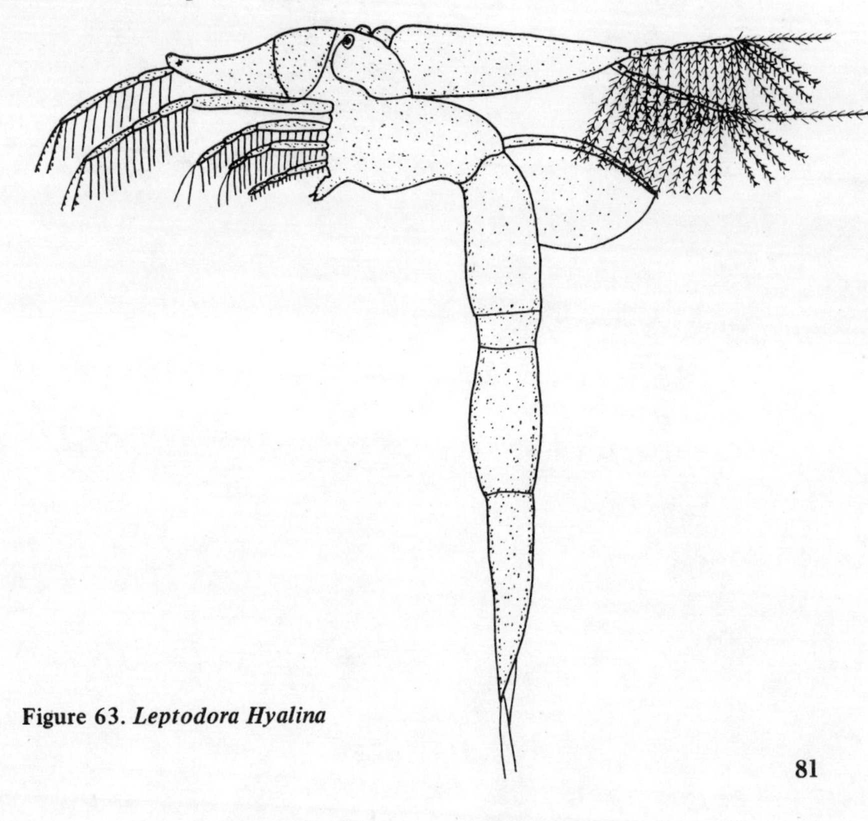

Figure 63. *Leptodora Hyalina*

One of the largest of the cladocerans is *Leptodora hyalina* (family Leptodoridae), which measures about three-quarters of an inch in length. It has a rudimentary shell that does not cover the legs or the long, segmented abdomen (Fig. 63), which ends in two claws. It inhabits freshwater lakes in America and in Europe and is a fierce carnivore, usually on the move in pursuit of smaller water fleas, rotifers, and protozoans. *Polyphemus pediculus* (family Polyphemidae) is also a carnivore. It has a highly colored transparent body and occurs in deep lakes and rivers throughout America and Europe (Fig. 64).

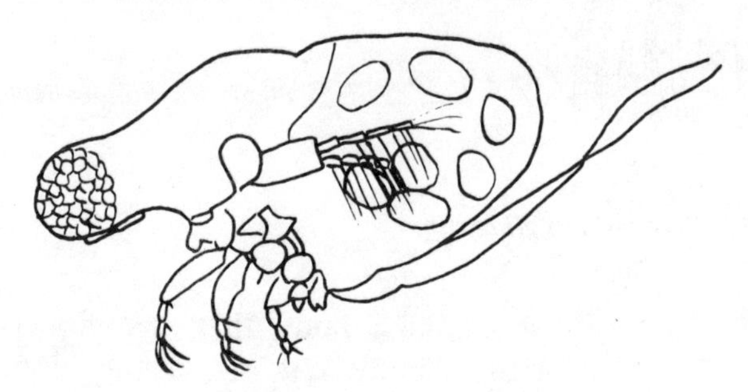

Figure 64. ***Polyphemus pediculus***

5
The Ostracods

Look at an ostracod and you will think you are looking at a tiny clam. For the ostracod has a two-valved shell, hinged at the back like a clam, and gaping open below through which the appendages are thrust out (Fig. 65).

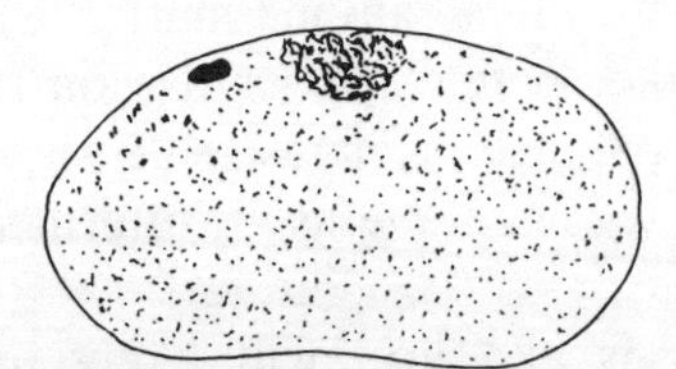

Figure 65. ***Eucypris virens***

The body of the ostracod lacks any segmentation and is laterally compressed. It is entirely enclosed within the two-valved carapace, the two valves being held together by a transverse muscle that permits the two valves to close and open at the ventral side as in the clam and essentially the same in function as the muscles we eat as fried scallops. Unlike that of the clam, however, the bivalve shell doesn't have any growth lines.

The ostracods have seven pairs of appendages: two pairs of antennae, which are used for both locomotion and orientation, a pair of mandibles, two pairs of maxillae, the second pair usually being leglike, and two pairs of legs, the second pair frequently being bent back and used to prevent dirt and other debris from getting into the shell. The abdomen is short and often ends in a projection with two terminal claws, called the furca. The ostracods have a single median eye or two eyes set close together and they usually lack a heart. Their internal organs are very compact, a distinguishing feature.

The ostracods, or seed shrimps as they are also known, are small animals, measuring from less than a sixteenth of an inch to about an eighth of an inch in length, though some marine forms may be as much as five-eighths of an inch in length. In color they may be white, yellow or gray-green, red, brown, or black; those which live among algae are often green. Most of them are marine, the freshwater forms being found in all types of fresh waters, though they are rarely found below three feet deep and few live in swift streams. Only occasionally may they be found in plankton. They are abundant in quiet, shallow waters, especially those filled with plants. They are for the most part creeping animals, crawling about in the algae or over vegetation or even burrowing in the mud and ooze on the bottom, though there are some that are free-swimming, but these are apt to be in brightly sunlit spots. One rare species clings upside down to the surface film in open water. They generally frequent the sunny protected areas of ponds and move along the bottom in a somewhat unsteady manner, and yet they are able to scurry quickly, changing their gait as the need arises or as the whim of the moment dictates. Their antennae are in almost constant motion.

The "seeds with legs" are essentially scavengers, feeding on decayed plant material as well as on dead animal matter. They also eat bacteria, molds, algae, and the like. A few species live on fish, amphibians, and crayfish, though not apparently as parasites. Fish seem to feed on them only sparingly. Several species serve as an intermediate host of certain fish tapeworms.

Collect some ostracods and you will likely find that they are all females, since males are generally present only in small numbers. Some species are parthenogenetic, there being no males or at least none have been found. All our North American species, with one exception, lay their eggs singly or in clumps or rows on plants, floating twigs, or rocks. The one exception produces eggs that hatch within the female's body, in other words the young are born alive. In some species the young are produced as nauplii, in others they are produced later than the nauplis stage. As a point of interest, some European ostracods have sperm cells that may be up to ten times as long as the males themselves and longer than the sperms of whales and elephants. It is unlikely, however, that these unusually long sperms are functional, since most of the eggs develop parthenogenetically.

In some species several generations of parthenogenetic females succeed one another until a generation of males and females is produced, the males and females mating and producing fertilized eggs that remain dormant throughout the winter; in others only parthenogenetic females are produced, males being unknown. You will find adult ostracods at any time of the year; the adults of some species are numerous in late winter, of others in early or late spring. During the winter you will find active ostracods in the mud beneath the ice of a pond, and should you scoop up some of the top silt from the bottom and let it stand in a warm place, the ostracods will usually appear on the surface.

An ostracod common everywhere in fresh water is *Cypridopsis vidua* (Fig. 66). It is yellowish-green with three transverse bands on the back and sides. Also often quite common is *Candona acuminata* (Fig. 67). This species is white or brown with its shell pointed at the posterior end. Both Cypridopsis and Candona belong to the family Cypridae. Also a member of this same family is the cosmopolitan species *Eucypris virens* (Fig. 65). It is bluish-black above and greenish on the sides with yellowish areas extending diagonally downward from around the eyes. A related species, *Eucypris fuscata* (Fig. 68), is also cosmopolitan in distribution. It has a kidney-shaped shell and is greenish-brown in color with transparent spots and a bluish-black patch on either side.

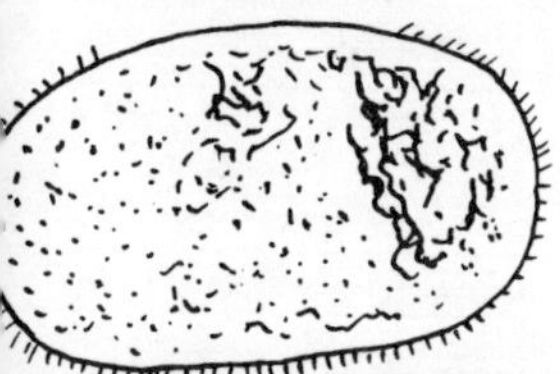

ure 66. ***Cypridopsis vidua***

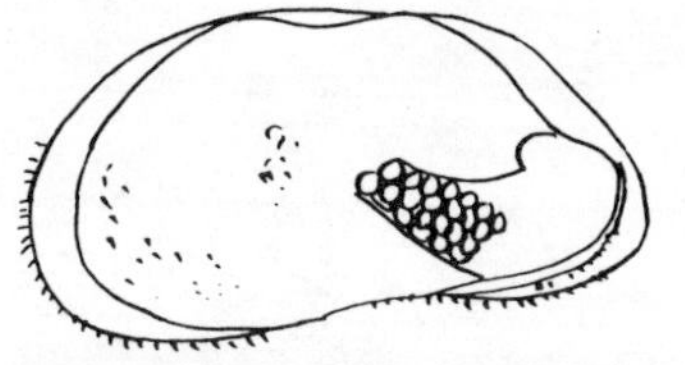

Figure 67. ***Candona acuminata***

Figure 68. ***Eucypris fuscata***

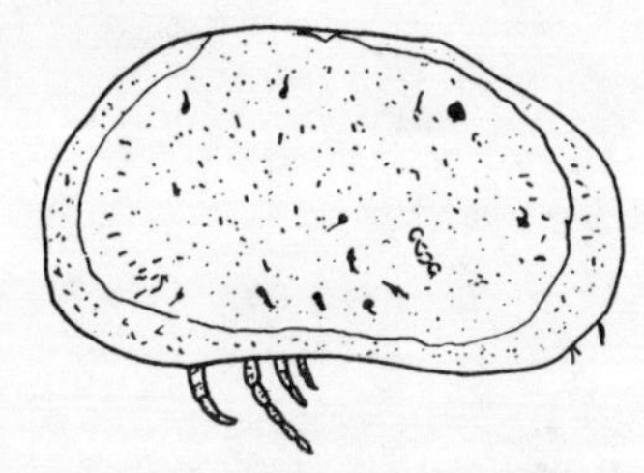

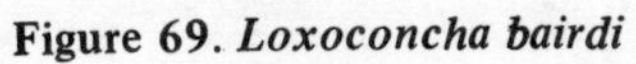

Figure 69. ***Loxoconcha bairdi***

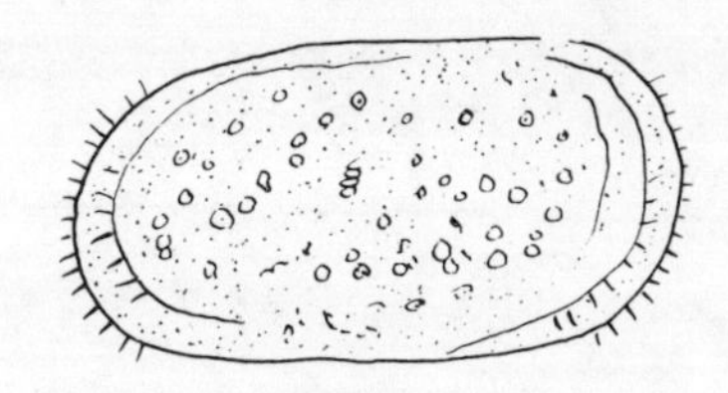

Figure 70. ***Cytheris arenicola***

The following three species, members of the family Cytheridae, are common in Vineyard Sound: *Loxoconcha bairdi* (Fig. 69), its shell with a notch of the dorsoposterior angle and its ends and ventral margin extended into a flattened rim, occurs among eel grass; *Cythereis arenicola* (Fig. 70), with a quadrangular shell, lives on sandy bottoms; and *Cytheretta edwardsii* (Fig. 71), the top and bottom margins of its shell nearly parallel, the rounded ends hairy, inhabits rather deep water. There are about seven hundred species in the family, all of them usually with a hard and calcareous shell, all of them exclusively marine, and all unable to swim.

Widely distributed in both the Atlantic and Pacific Oceans are the two species *Halocypris brevirostris* and *Conchoecia magna* (family Halocypridae). They both have a very thin and flexible shell with a notch in front, the antennal sinus, for the protrusion of the second antenna, above which is a rostrum. In *brevirostris,* the rostrum is short, in *magna* it is well developed.

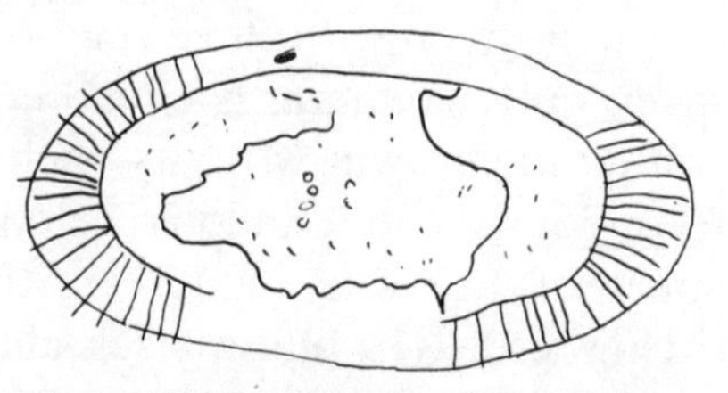

Figure 71. *Cytheretta edwardsi,* Male

6
The Copepods

You can most easily recognize a copepod by its cylindrical shape, its greatly narrowed abdomen, and by the pair of projections that extend from the tip of the abdomen (furca) and that are armed with small spines or setae (Fig. 72). The copepods have legs that are flattened like the blades of an oar, hence their name, the word *copepod* coming from two Greek words meaning oar and foot, in other words "oarfoot." There are a great many species of copepods–some authorities put the number at more than six thousand–and all are small, many microscopic in size, the largest less than half an inch in length.

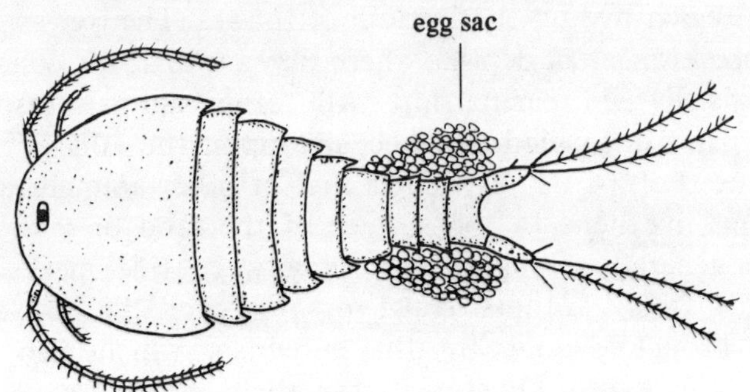

Figure 72. Cyclops, Female *(Cyclops viridis)*

Like all crustaceans their bodies are more or less distinctly segmented, with the exception of some parasitic species, and are typically divided into head, thorax, and abdomen, though in some cases the head and thorax have become fused together to form a cephalothorax that may be completely or partly protected by a carapace. The head, as a rule, bears six pairs of appendages: two pairs of antennae, one pair of mandibles, two pairs of maxillae, and a pair of maxillipeds. The first pair of antennae (antennules) are uniramous and are primarily sense organs but may be used for locomotion, for suspension, or even, in some species, by the males as clasping organs. The second antennae may be either biramous or uniramous and are often employed as grasping organs. The mandibles, maxillae, and maxillipeds are normally well developed and functional but in some species they may be much reduced or even absent.

The thorax consists of six segments, each with a pair of legs. The first four pairs are branched, and being flattened and paddle-shaped are used for swimming. The fifth pair may be like the first four or slightly modified or absent altogether; when present the legs may be modified into grasping organs in the male. The structure of this fifth pair is a characteristic much used in the identification of species as well as of genera. The female lacks the sixth pair of legs; the male does, too, in many instances but when present is usually rudimentary.

The sixth segment of the thorax, though considered by some authorities to be a part of the abdomen, is generally called the genital segment, since it contains the genital organs. The segments of the abdomen vary from one to six, none of which bears any true appendages. There is at the terminal end, as I have already mentioned, a pair of projections that are provided with bristles, spines, or setae.

The copepods are widely distributed, and though most of them are marine, there are many freshwater species; there is hardly a pond to be found without some of them. They include free-swimming forms, commensals, and parasites, and also what are known as benthonic animals, those which live on the bottom of the sea. The free-swimmers occur in the plankton at all depths, where they move about constantly; the commensals live in partnership with other animals, especially ascidians; the parasites, called fish lice, live upon the gills or skin or burrow into the flesh of fish as well as that of other animals; and the benthonic forms live beneath the surface of the sand or mud on the bottom, more generally along the shore between tide marks. Some copepods swim freely, others crawl or run over the substratum. Swimming is brought about by the antennae, which also aid in equilibrium and in depth adjustment, but their jerky forward movements so typical of many of them are effected by the backward movements of their hind legs.

The distribution of the copepods is largely affected or influenced by temperature. Thus many species prefer the cold of arctic and deep-water lakes; others the warmer parts of small ponds or pools.

Most copepods are whitish, brownish, or grayish in color. Species that live in shallow water where light reaches the bottom are inclined to be brighter, and in the spring some species that live along the weedy margins of ponds may often be a pink, red, green, or blue. Copepods that live in deep water are apt to be translucent, as well as smaller and more slender. The bright reddish color often seen in Diaptomus is due to the oil globules stored within the body.

The copepods eat, so it seems, whatever they are able to swallow, but most of them feed primarily on organic detritus, on microscopic organisms and any kind of decayed matter that they scrape or rake from the bottom. Some, however, feed only on plankton that is filtered out by the mouthparts from the currents of water that the waving of the antennae bring to them. Those parasitic on fish obtain their nourishment from the tissues of their host. Usually the parasitic species are not present in sufficient numbers to pose any serious threat, though at times they can inflict considerable damage, especially in fish hatcheries. Perhaps more important are the species that serve as intermediate hosts of tapeworms, flukes, and nematodes. In Africa, Asia, and the West Indies a species of cyclops is the intermediate host of the guinea worm, a serious parasite of man, and in our own country species of cyclops and diaptomus are intermediate hosts of the fish tapeworm that may infest man.

It seems to be a matter of debate just how important a role the copepods play in the aquatic food chain. Some authorities claim that the role is an insignificant one; others that it is most important. There is no question but that they provide a link between the inorganic constituents of the water and the food supply of many of the animals that live in it, especially since they occur in immense numbers. Young fish feed on them for a time and such marine fishes as mackerel, herring, sardines, and smelts continue to feed on them throughout their lives. Certain whales also take them as food.

Anyone who might watch the copepods in a pond would likely observe mating pairs swimming together. The males transfer their sperm cells to the females in small packets, and the eggs may be fertilized within a few minutes or not for several weeks. The females carry the eggs on one or two sacs (Fig. 72) attached to the ventral side of their bodies, and the nauplii that hatch from them may become adults within a week or not for several months or even a year.

Doubtless the best known of the copepods is Cyclops, named, because of its single eye in the middle of the eye, after the mythological race of one-eyed giants who, as Homer tells us, were extremely troublesome to Ulysses on his way back from the Trojan War. You can see it in almost any pond as a tiny white speck moving jerkily through the water.

Cyclops has a pear-shaped body (Fig. 72) tapering into a forked tail, with the first antennae very long and outspread like antlers, and four pairs of swimming legs on the thorax. There are several species of Cyclops (family Cyclopidae), the most common one found in small ponds throughout the country being *Cyclops viridis.* It is extremely variable in color but is usually greenish. Structurally the first antennae have seventeen segments, the second four; the first four pairs of legs are three-segmented and biramous; the fifth pair two-segmented and uniramous, the end segment with a long feathery bristle.

During the mating process of Cyclops, the male clasps the female with his antennae and they swim around together for long periods sufficient for the male to transfer enough sperms to the female to fertilize several clutches of eggs. The female starts laying her eggs shortly after mating, the eggs emerging from two oviducts and being fertilized as they pass a seminal receptacle into which the sperm cells had been deposited during mating. The eggs are held together by a gelatinous secretion in two ovisacs that are attached to the sides of the genital segment. A female may produce, during her lifetime, up to thirteen pairs of ovisacs, each consisting of fifty eggs. It has been calculated that a single female would give rise to three million offspring during each year if all of the many generations survived.

The eggs hatch into nauplii, which undergo eleven molts before becoming adults. Like other copepods Cyclops produces two kinds of eggs–summer ones that form in rapid succession and develop quickly and winter or resting eggs that can endure cold and drought and that remain dormant even through years of drying.

The copepods that are free-living pass through periods of activity and inactivity; thus a cold-water species will have one generation rapidly succeeding another and large numbers will swarm beneath the ice or in deep, cold waters while in warmer waters the species will exist only as resting eggs. On the other hand, warm-water species will do just the opposite, in other words males and females will be numerous in the summer while their resting eggs will lie dormant during the winter. Some species of copepods may go through several cycles of active animals and resting eggs each year or may have generations all the year and be more or less perennially active.

As I have already remarked, the copepods are greatly influenced by temperature. Thus adults of a cold-water species may be active only during the winter in one body of water, but in another where the temperature may be lower they are active throughout the year. In large lakes the copepods live largely in the upper waters, but many of them migrate short distances as the temperature changes. As a rule they live in summer chiefly near the top, where the surface water is warm, but in winter when the surface water becomes cold they descend to warmer, deeper water. In short, copepods shift vertically not only with seasonal temperature changes but also with changes of day and night and even from hour to hour. And light, too, is a factor in determining their movements. Copepods also seem to appear to prefer different environments according to age. Thus nauplii of open water species are likely to be close to the surface in brilliant light while the adults occur at slightly lower levels.

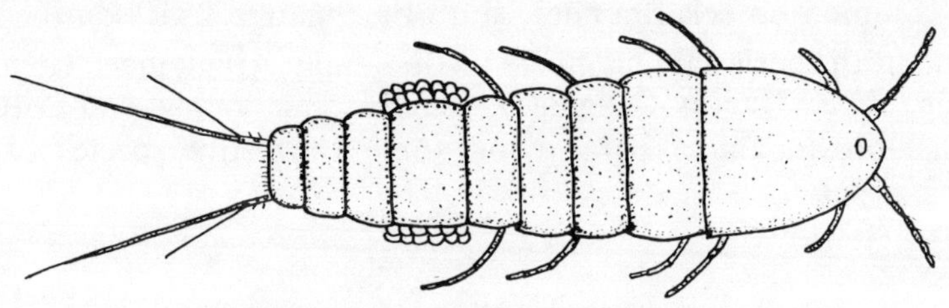

Figure 73. ***Canthocamptus minutus***

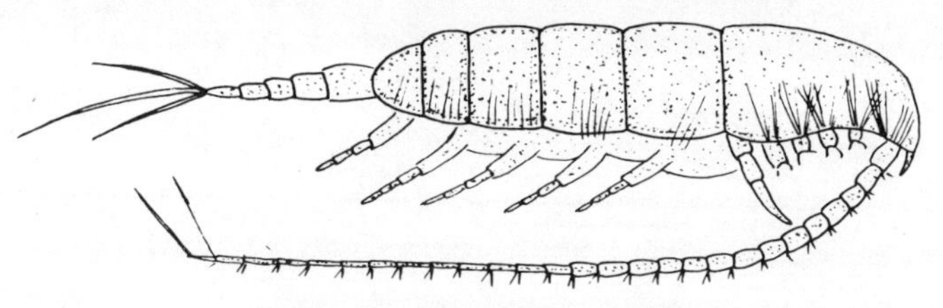

Figure 74. ***Calanus finmarchicus***

There are many species of copepods and lack of space limits me to mention only a few, such as the following: *Diaptomus sanguineus* (family Diaptomidae), which has a bright red body and is quite common, though it occurs only in early spring pools *Canthocampus minutus* (family Canthocamptidae), hyaline in color and common in muddy pools in the eastern United States and in Europe (Fig. 73); *Calanus finmarchicus* (family Calanidae), an unusually large species reddish or yellowish in color and found in the sea, where it is sometimes so abundant as to color large areas of the sea red or yellow (Fig. 74); and *Centropages typicus* (family Centropagidae), another marine species, reddish or bluish in color, very abundant in the Gulf Stream south of Martha's Vineyard Island and quite common on the New

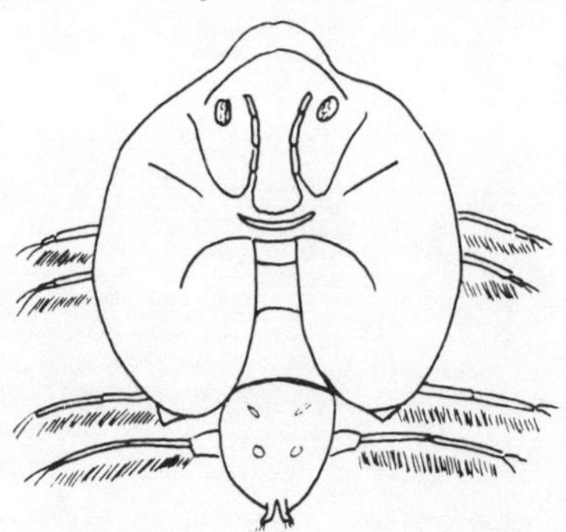

Figure 75. ***Argulus versicolor***

England Coast. I might also mention such parasitic species as *Argulus versicolor* (Fig. 75), which is brilliantly variegated with red, green, and orange and is common on pike and pickerel in the New England lakes, and *Argulus laticauda,* which is black, mottled with yellow and brown and common on eel, flounder, and other marine fishes on the Atlantic coast. Both species are members of the family Arguloidae. Then there is *Caligus rapax* (family Caligidae), which is pale yellow and dotted with reddish-brown and a parasite on some thirty-five species of marine fishes.

7
The Amphipods

Walk along the sea beach (of the Atlantic coast) as I have often done and you will be sure to see the beach fleas, small brown or dark-colored animals that live beneath the decaying seaweed and when disturbed hop and run with great rapidity. They look like fleas with their arched backs and narrow bodies, but they are not fleas, for fleas are insects and the beach fleas are crustaceans, more specifically, amphipods, but they look like fleas and so are called beach fleas.

The amphipods have an elongated body that is usually compressed from side to side, and they have no carapace. The first segment of the thorax (in some species the second also) is fused with the head and the remaining seven are free-jointed, that is, they appear as distinct segments. There are six segments in the abdomen, though sometimes there are less, and the appendages are well developed. When all the appendages are present there are the two pairs of antennae, one pair of mandibles, two pairs of maxillae, and one pair of maxillipeds (which are attached to the first segment of the thorax), seven pairs of periopods, the first two of which are called gnathopods and which are used for grasping the food, and six pairs of pleopods or abdominal appendages. The gnathopods are usually larger than the others and are provided with pinching claws. The first three pairs of pleopods are adapted for swimming, the last three pairs are short, stiff, and modified or jumping, these last three pairs sometimes being called first, second, and third uropods. The eyes are sessile, that is, they are not set on movable stalks, and the gills are attached to the thoracic legs.

In color the amphipods vary greatly. Many species are colored crimson, red-brown, bright green, and deep blue-green. Those found on beaches are usually in shades of gray, white, or are simply translucent.

During the mating process the males transfer their sperm cells to the females where they fertilize the eggs that have been placed in a brood pouch (marsupium) and that may number from two to two hundred or more. In some species the males carry the females on their backs during mating, swimming about in such a position for sometimes as long as a week. Unlike most crustaceans the amphipods do not have a larval stage. The eggs hatch in nine to thirty days as miniature adults, the newly hatched young remaining in the brood pouch for several days. The females may have several broods a season, but a new mating is necessary in each case.

There is said to be over three thousand species of amphipods, the vast majority of which are marine, living in algae that literally swarm with them or on the bottom of the coastal shelves. Many species live on the sea beach, burrowing in the sand or living under stones or decaying vegetation. A few species occur in fresh water but some of these are well known because they are one of the favorite food of fishes such as the brook trout, which eat them in enormous numbers. These species live in water where plant life is abundant, about water cress, in tangles of Chara, and beneath decaying leaves. The amphipods thrive in cold water and are active throughout the year, hiding in vegetation or under debris in both quiet and running waters. Except for the deep lake species they are generally not found more than three feet below the surface. There are a number of blind species of caves and underground springs, and these may sometimes appear in woodland seepages. Marine amphipods occur in great numbers in arctic waters and the ill-fated Greeley expedition escaped from starving by feeding on them.

The amphipods are agile water acrobats and swim, jump, climb, or glide with equal facility. The freshwater forms seem to be most active at night, walking or crawling about by flexing and extending the entire body. When swimming they hold their bodies out straight.

Although some pelagic species are carnivorous, most of the amphipods are omnivorous scavengers, feeding for the most part on dead animal and plant material and browsing over aquatic vegetation for the microscopic film of organic material usually present there. Some, however, are herbivorous and a number are parasitic. In turn they serve as food for other animals such as fish, birds, insects, and amphibians, many forming a large part of the diet of commercial and game fish. The stomachs of whales and seals have been found to be gorged with them. One species, *Chelura terebrans,* bores into submerged wood and timbers and often does considerable damage.

Beach Fleas The beach flea I have mentioned is *Orchestia agilis* (Fig. 76). It occurs in countless numbers beneath the masses of sea wrack on the beach and ranges along the Atlantic coast of North America and Europe. There are about twenty-five species of Orchestia (family

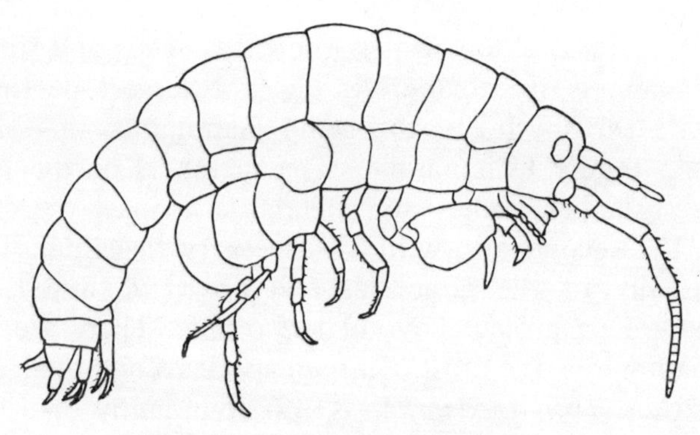

Figure 76. *Orchestia agilis,* Male

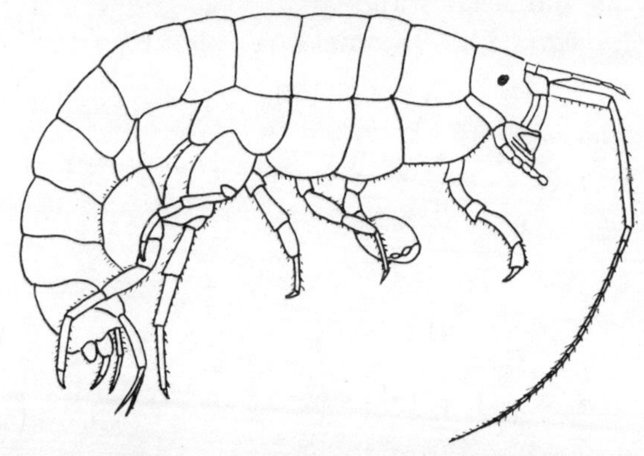

Figure 77. *Talorchestia longicornis,* Male

Orchestiidae) and all are dark colored with the first antennae shorter than the second. *Orchestia agilis* is brown in color to match the seaweeds among which it lives and so, too, is *Orchestia palustris,* which is a little larger than *agilis,* the latter measuring about half an inch. There are minor structural differences that serve to distinguish the species. *Orchestia palustris* does not live in association with agilis but is usually found in salt marshes some distance from the beach, where it lives among the grass and weeds.

Common on the sandy beaches around the high-water mark, where it burrows in the moist sand during the daytime, is the species *Talorchestia longicornis* (Fig. 77), of the same family. It is a little larger than *Orchestia agilis,* measuring about an inch in length, with large, conspicuous eyes, and is paler in color, being almost white or at least whitish. Sometimes, however, it is brown. A similar species, *Talorchestia megalopthalma,* has even larger eyes that cover most of the sides of the head. It also has shorter antennae. Both species are found together and range from Cape Cod to New Jersey.

Scuds and Sideswimmers I should, perhaps, mention at this point that the amphipods are also sometimes known as scuds and sideswimmers from their manner of moving. Among the largest of the scuds or amphipods are the species of the genus Gammarus, one of over fifty genera of the family Gammaridae. The members of the family have comparatively slender bodies and slender antennae, the first usually longer than the second and with an accessory flagellum. The gnathopods are usually of the same size and powerful, and the terminal pleopods extend or project beyond the others. There are over thirty species of Gammarus, the most common species along the New England coast being *Gammarus locusta.* It is also abundantly distributed from the Arctic Ocean along the coast of Europe and in the Pacific Ocean, both in American and Asian waters.

Gammarus locusta (Fig. 78) measures about an inch long, though

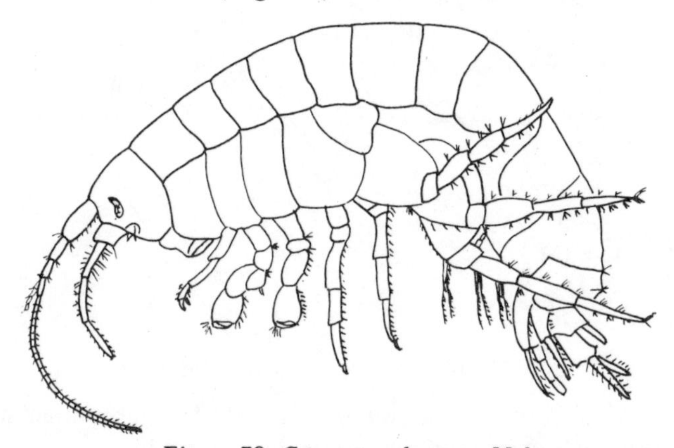

Figure 78. *Gammarus locusta,* Male

some individuals may reach a length of two inches, especially in the Arctic, and in color is olive brown to reddish. It is abundant under stones and seaweeds at low-water mark and also occurs down as far as fifty fathoms. This species and others of the genus do not jump like the beach fleas Orchestia but move rapidly, lying on the side, and in water swim with their backs downward. *Gammarus annulatus* occurs in the same places as the preceding but usually a little higher up on the beach. It is somewhat smaller and lighter in color and has dark bands with red spots on the sides of the abdomen. Annulatus is a surface swimmer and may often be found in tide pools as well as to thirteen fathoms in depth. Also associated with seaweed or under stones near the low-water mark is *Melita nitida* of the same family. It is about two-fifths of an inch long and is slate-colored or dark green.

A freshwater representative of the genus Gammarus is *Gammarus fasciatus,* which is common in freshwater ponds and streams of the eastern states. It measures about five-eighths of an inch in length, is whitish in color, and has both the first and second antennae of the same length. Another freshwater amphipod is *Hyallela knickerbockeri* (Fig. 79) of the same family as Orchestia. This species and *Gammarus fasciatus* are the most common amphipods in the ponds of all the eastern states. In early spring both gather in mats of the alga, Spirogyra, feeding on the dead filaments, though both have been observed eating freshly killed snails, tadpoles, and crushed members of their own species.

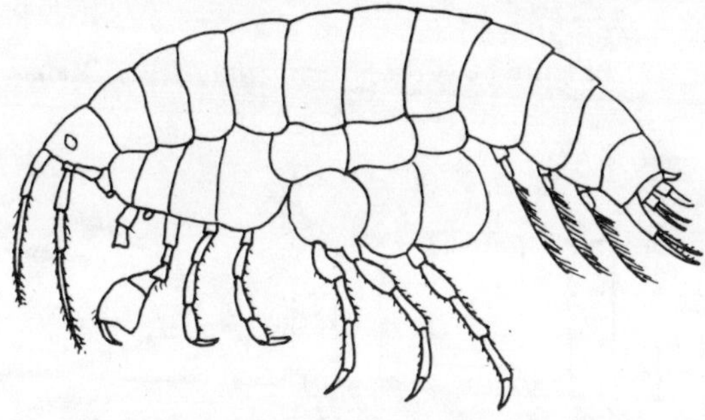

Figure 79. ***Hyalella knickerbockeri***

Although both these scuds or amphipods may be seen with the naked eye in their natural habitats, they can be better observed in an aquarium planted with a few water plants and supplied with a few dead leaves, for here, at close range with the aid of a reading glass, you can watch them perform all their acrobatic stunts and observe their feeding and mating habits very easily. When mating the male clasps the female with his grasping legs and swims about holding her beneath him for a period of several days. It is not unusual for a female to carry a previous brood of young in her thoracic brood pouch during the mating swim. Although the scuds are not as prolific as other crustaceans, one pair of Hyalellas have been known to produce a clutch of about a dozen and a half eggs fifteen times in one hundred and fifty-two days.

Members of the family Corophiidae are tube dwellers. They have a depressed body and a small abdomen with the fifth pairs of periopods being the longest. A common species found along the entire coast of southern New England and on the New Jersey shore is *Corophium cylindricum* (Fig. 80). It measures about an eighth of an inch long, is purplish brown with a dark crossband at the posterior margin of each segment, and lives in soft tubes that are a familiar sight about the base of eelgrass. Another species common on sandy or rocky bottoms from Labrador to New Jersey is *Unicola irrorata* (Fig. 81). It is about three-fifths of an inch long, is brightly colored with variegated crimson blotches and a lengthwise median band of crimson spots on either side, and often lives in tubes not of its own making such as those of the annulate worms.

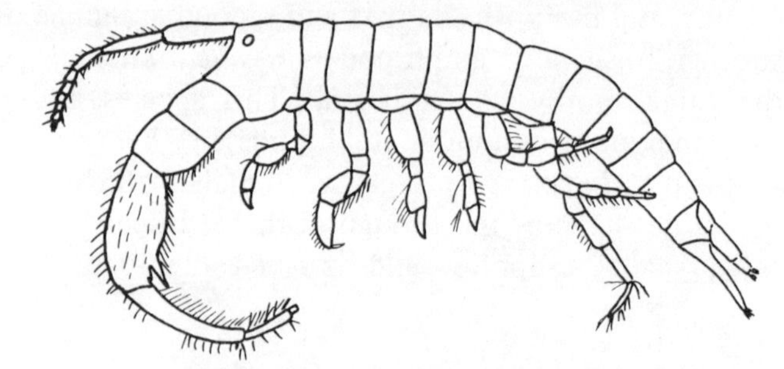

Figure 80. *Corophium cylindricum,* Male

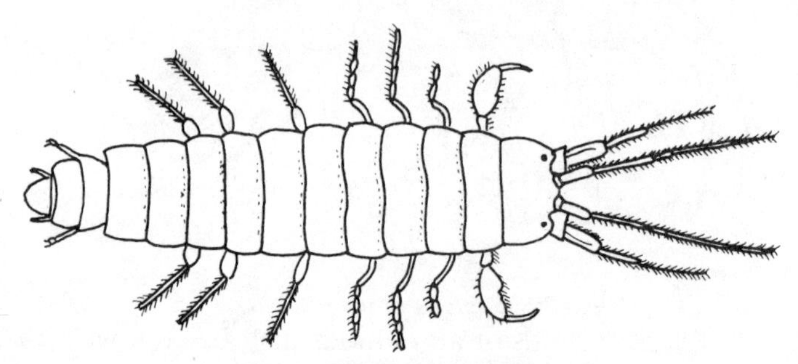

Figure 81. *Unicola irrorata*

I have already mentioned *Chelura terebrans* (family Cheluridae), the amphipod that bores into wood and thus generally is referred to as the boring amphipod. It has a cylindrical body, the second antennae longer than the first, and with three pairs of caudal stylets (Fig. 82). In color it is semitranslucent, thickly mottled with pink. It is a very active and destructive amphipod, and is usually associated in its work with the isopod *Limnoria lignorum.* Unlike Limnoria, which makes narrow and cylindrical burrows in submerged wood, Chelura excavates larger burrows in oblique lines near the surface so that the wood appears as having been plowed.

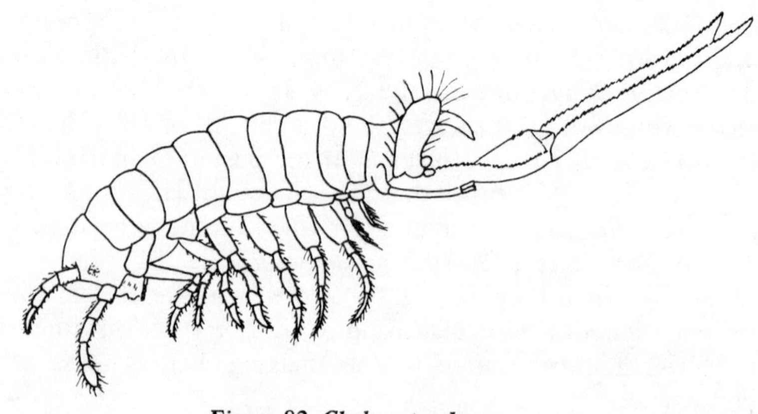

Figure 82. *Chelura terebrans*

There are several families whose members are burrowing in habits, such as Ampeliscidae and Haustoriidae. The species of Ampeliscidae have a head that is almost square in front, have usually four eyes, have the second antennae longer than the first, and a telson that has a deep median cleft. *Ampelisca macrocephala* is a common species found from the Arctic to Long Island Sound and in Europe. It is about five-eighths of an inch long, is whitish in color, and occurs, for the most part, among eelgrass where it burrows in the muddy bottoms.

The members of Haustoriidae are said to be the fastest of the burrowing amphipods, but just how much more quickly they can burrow compared to others I don't know. The periopods of these animals are very broad and modified for digging, a feature that serves to identify them. The species *Haustorius arenarius* is representative of the family. It is found on the seabeach near high-water mark from Cape Cod to Georgia and also in Europe. It is about three-quarters of an inch long, is whitish or pale yellow in color, closely resembling its sandy environment, and has dense plumes of setae on its appendages (Fig. 83).

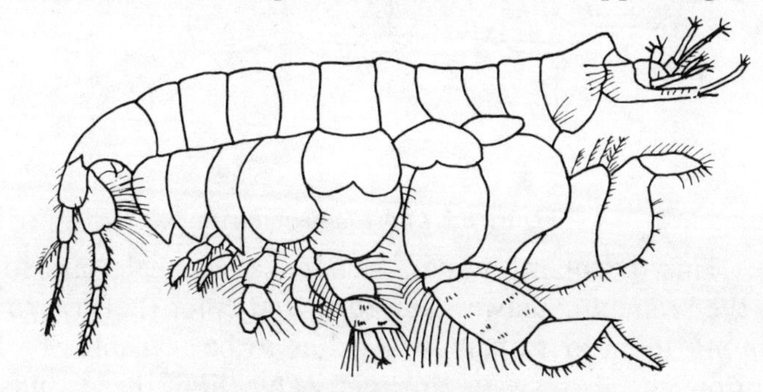

Figure 83. ***Haustorius arenarius,*** **Female**

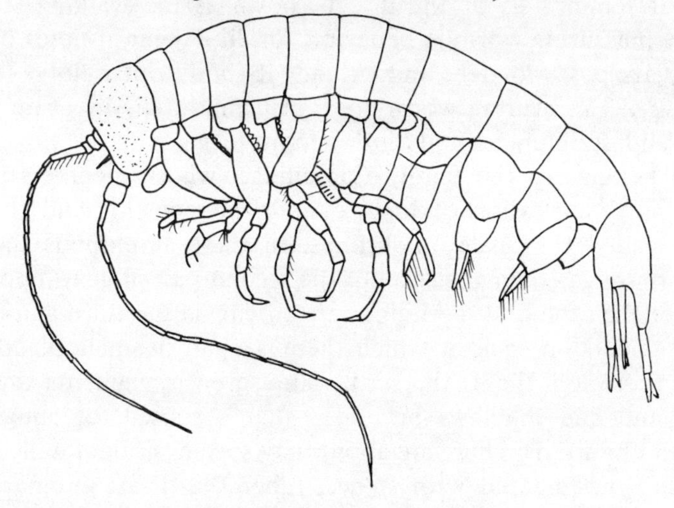

Figure 84. ***Hyperia galba,*** **Male**

The family Hyperiidae is a parasitic one, that is, its members are parasites on pelagic animals, especially jellyfish. They have a large head that is almost entirely occupied by their enormous eyes. The species *Hyperia galba* has an unusually high arched back and kidney-shaped compound eyes. It measures about three-fifths of an inch in length and is abundantly distributed from northern Europe to the Arctic Ocean and from Greenland and Nova Scotia down along the entire coast of New England (Fig. 84). It is parasitic on the white jellyfish, *Aurelia aurita.* Another species, *Hyperia medusarum,* common along the coast of New England, is found in the pink jellyfish, *Cyanea arctica,* as well as in other species of jellyfishes.

The Skeleton Shrimps One of the oddest animals to be found in the rocky tide pools is the little caprella and a glance at Figure 85 will show

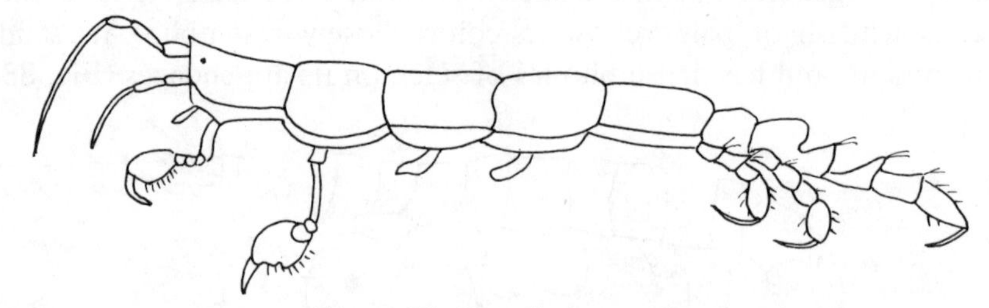

Figure 85. ***Caprella geometrica***

you why. This peculiar animal, slender as a skeleton, so closely resembles the branching seaweeds in form and color that it is practically hidden among them or so camouflaged as to be invisible. To heighten the deception, it also sways in the water like the fronds of the seaweeds. It reminds us of the insect known as the walking stick in form or the measuring worm in behavior, for, like them, it holds on to a support by its posterior feet and extends its body out rigidly. It even walks the way a measuring worm does, bringing its hind feet up to the front ones and doubling its body into a loop.

Caprella belongs to the family Caprellidae, whose members have a body that is a succession of long, cylindrical segments and that are sometimes called the skeleton shrimps. In these amphipods the first antennae are longer than the second, the second pair of gnathopods are much longer than the first, periopods are absent on the third and fourth thoracic segments on each of which there is a pair of small, bladderlike gills, and those on the fifth, sixth, and seventh segments are well developed and end in claws or hooks that are used for clinging to seaweeds and hydroids. There are about fifty species in the family, some of which are ornamented with spines, tubercles, or a combination of these characteristics.

Caprella acutifrons, whose color is variable, is a common species with a cosmopolitan distribution. It occurs on our Atlantic coast from Maine southwards. *Caprella linearis* is found on both sides of the Atlantic, on the American side being abundant from Greenland and Labrador to Long Island Sound, and *Caprella geometrica* (Fig. 85), also of variable color, ranges from Cape Cod to North Carolina. It may be found among eelgrass, in sponges, and on wharf piles.

Whale Lice Closely related to the Caprellidae are members of the family Cyamidae, which are known as the whale lice, since they live externally on the skin of these mammals. They are whitish, flat creatures with strongly indented segments, each one prolonged laterally and each joined to one another by contracted articulations, the abdomen being reduced to a rudimentary knob. The whale lice have a single pair of antennae, two pairs of gnathopods, the second being the larger with an unusual kind of claw, with the last three pairs of periopods enlarged and strongly clawed. The segments between those bearing the gnathopods and the periopods do not bear periopods but instead each is equipped with a pair of long, fingerlike breathing organs or gills. The species *Paracyamus boopis* (Fig. 86) is typical of the family.

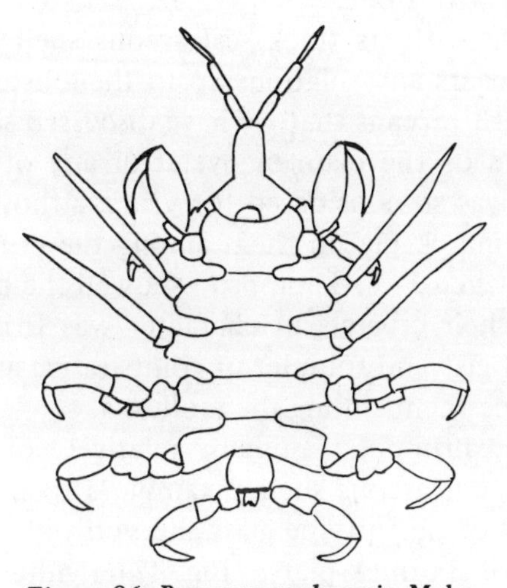

Figure 86. ***Paracyamus boopis,*** **Male**

8
The Cirripeds

Who has not heard of the barnacles? Everyone who has sailed the seas knows these animals and how they attach themselves to the hulls of ships and impede their speed. Along the North Atlantic coast they are a familiar sight, where they are attached to the rocks in countless numbers and so whitening them that from a distance they look as if they had been dusted with flour.

Look at a barnacle with its thick, calcareous shell and you would hardly think that it bears any relationship to the lobster, crab, shrimp, or any of the other crustaceans that we have discussed so far. It appears more to be a relative of the clam or oyster or any other mollusk for that matter. And it was so considered for years, although in 1802 the French naturalist Lamarck placed them among the crustaceans. But it was not until 1830, when J. V. Thompson showed the barnacle larva to be a nauplius, that their crustacean character was firmly established. Yet the French zoologist and founder of comparative anatomy, Cuvier, still continued to place them among the mollusks.

The barnacles are shrimplike animals, relatively of large size, and usually enclosed in a calcareous shell. Examine a rock barnacle if you will with a hand lens or magnifying glass and you will see that the shell is usually composed of six thick plates (Fig. 87), rigidly attached to one

Figure 87. *Balanus balanus*

another and to a fold of skin surrounding the body. At the top are two valves that open and close like double doors, and it is through these doors that the appendages, fringed like feathers (Fig. 88), are extended when the animal is covered with water. They sweep through the water like a casting net in a regular rhythm, beating back and forth as many as a hundred times a minute, capturing small, swimming creatures and organic matter that serve as food. As Thomas Huxley once remarked, "A barnacle is a little shrimplike animal standing on its head within a limestone house and kicking its food into its mouth with its feet."

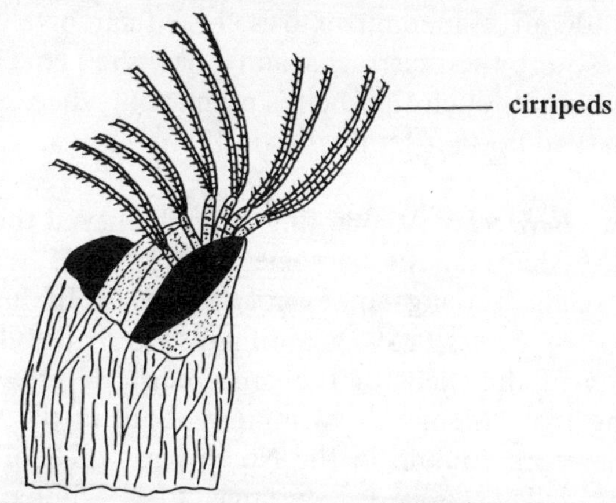

Figure 88. Rock Barnacle with Appendages Extended

The appendages of the barnacles are quite characteristic and consist of a pair of mandibles, two pairs of maxillae, and six pairs of biramose and feathery thoracic appendages or legs. It is these last which project from the shell and give the animal its characteristic appearance, also the name to the group of which the barnacles form a part: Cirripedia, which is from two Latin words, *cirrus,* plural *cirri,* meaning curl or ringlet, and *pes* or *pedis,* meaning foot (Fig. 88).

The abdomen of the barnacles is rudimentary. They have a food canal that passes straight to the anus at the posterior end of the abdomen, a digestive gland, and excretory tubes. They also have a nervous system, which consists of a brain and a chain of five or more ganglia, but no respiratory or circulatory systems, the cirripeds or feet performing these functions.

In a number of species the barnacles are unisexual but most of them are hermaphroditic, that is, they have both sexes, a condition doubtless developed by their sessile mode of living, though in a few genera complementary males do occur that live in or near the genital opening of the hermaphroditic individuals. In the species where the sexes are separate, the males are minute animals and consist of little but genital organs; they live a parasitic life on the females. Young barnacles emerge from the eggs as nauplii. They are free and independent animals with one eye, three pairs of legs, and a single shell. They swim around for a while and molt several times when they develop two eyes, two shells, and six pairs of legs. This later larval stage is known as the cypris stage from the animals' external similarity to an ostracod. At this time they attach themselves to some object that will be their permanent home, fastening themselves to the object that they have selected by means of their antennae, which have become suckers, and making their hold secure by secreting a cement to hold them firmly in place.

They now undergo a metamorphosis, lose their bivalve shells and their eyes, and acquire their cirripeds, and a new shell covering. Further growth is by molting, though the shell is permanent, successive stages of growth being marked upon it by lines as in the mollusks.

Stalked or Goose Barnacles At one time it was believed that geese were hatched from the shells of the barnacles that we today call the goose barnacles, the shells looking somewhat like eggs. In his *Herball or Generale Historie of Plants* (1597), the English herbalist Gerard tells us in his quaint way of the shells of the goose barnacle growing on a tree and geese falling from them and swimming about in the water below. He writes: "There are founde in the North parts of Scotland and the islands adjacent called Orchades certaine trees whereon do growe certaine shell fishes of a white color, tending to russet, wherein are conteined little living creatures; which shells in time of maturitie do open, and out of them grow these little living foules whom we call barnakles, in the North of England brant geise, and in Lancashire tree geise; but the other that do fall upon the land do perish and come to nothing." After describing the various transformations he goes on to say: "But what our eies have seen and hands have touched we shall declare."

The goose barnacles belong to the genus Lepas (family Lepadidae). They have an elongated, flattened body enclosed in a shell consisting of two valves each of which is composed of a number of large calcareous plates that usually do not number more than five. The body is mounted on a more or less slender, fleshy stalk known as the peduncle and the part of the body enclosed by the shell is called the capitulum. The shell consists of five plates, and these are a pair of fan-shaped scuta that cover the stalk end of the body, a pair of smaller terga at the opposite end and covering the upper end of the body, and a median dorsal carina or keel at the hinge of the double shell (Fig. 89). Because of the shape of the shell and stalk these crustaceans are known as the gooseneck barnacles or simply goose barnacles.

A cosmopolitan species is *Lepas fascicularis,* which is often numer-

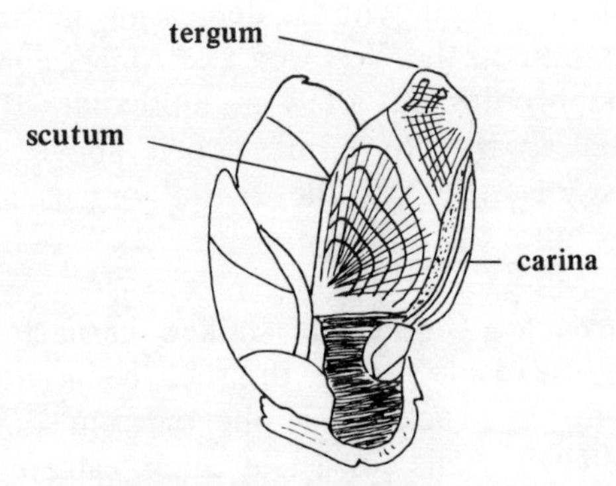

Figure 89. *Lepas anatifera*

ous, especially in early summer, on the North American coast, ranging from the Bay of Fundy to Florida. It is also found along the Pacific coast north of San Francisco and is usually found attached to seaweed and other floating objects. Its shell measures about two inches long, the plates being very thin and paperlike, with the carina bent at right angles. A second species is *Lepas anatifera* (Fig. 89), which is perhaps the most common species. Its shell is one inch long, bluish-white, the plates faintly striated. This species, found on ships' bottoms and floating objects, is also cosmopolitan in range and occurs more southerly than *Lepas fascicularis.* A third species, *Lepas hilli* (Fig. 90), also cosmopolitan in range but more northerly than the preceding, occurring south to Vineyard Sound and San Francisco, has a smooth, bluish white shell, which is about two inches or less in length, and the summit of the stalk pale or orange colored. The members of the genus are also commonly known as the ship barnacles, since they attach themselves to ships as well as floating logs. They are actually wanderers of the deep and multiply in such numbers on the hulls of ships as to seriously impede their progress. Aside from reducing their speed they do no injury to the ships.

There are other stalked barnacles as the members of the family Scalpellidae, for instance. These barnacles have a very long stalk, with scales or spines, and a shell that consists of a large number of plates, in one species, *Mitella polymerus,* as many as one hundred and eighty;

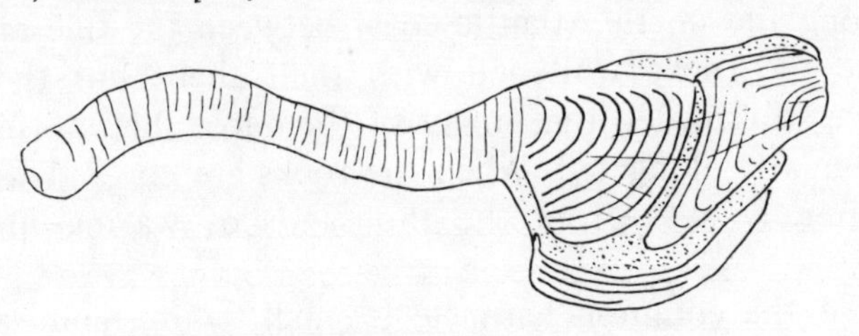

Figure 90. *Lepas hilli*

which are arranged in several whorls, decreasing in size from above downward. It is common on the West Coast of America. Most species of the family are hermaphroditic but a few are unisexual with complementary males, the males, as I have mentioned above, being minute animals that consist of little but the genital organs and that live a parasitic life on the bodies of the females.

Rock or Acorn Barnacles Unlike the stalked barnacles, the rock or acorn barnacles (Fig. 87) lack a stalk, the shell being attached directly to some solid substance as a rock, wharf pile, or, as in some species, to a sea turtle or whale. Their shells consist of thick, calcareous plates that fit together in a tentlike form or in the form of a cylinder, the plates usually being six in number with overlapping edges and are an unpaired carina, an unpaired rostrum, and two lateral pairs. The two pairs of lateral plates have their ventral edges overlapping and their dorsal edges underlapping the neighboring plates. The edges that overlap are called radii, those which underlap, alae. Within this cylindrical shell the animal lies on its back with the six pairs of thoracic feet uppermost, the legs being thrust through (Fig. 88) and withdrawn through (Fig. 91) the opening of the cylinder, which may be closed and opened by means of two pairs of hinged plates, called the opercular valves and which correspond to the scuta and terga of Lepas. Tap a rock incrusted with barnacles with you ear held close and you will hear the closing of many doors.

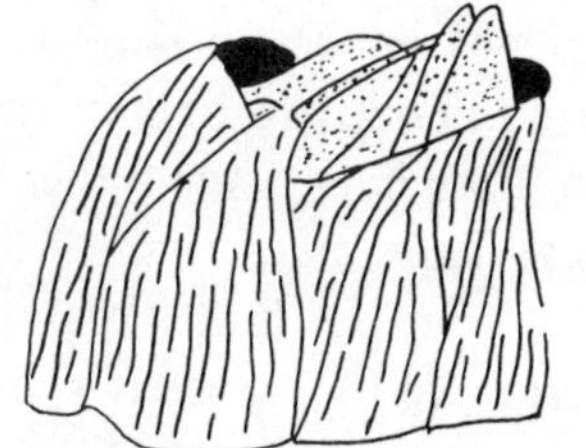

Figure 91. Rock Barnacle with Appendages Withdrawn

Figure 92. *Balanus eburneus*

The species that is generally known as the common barnacle is *Balanus balanoides* (family Balanidae). It is so enormously abundant on the rocks along the north Atlantic coast between the tide marks that not only are the rocks whitened with their shells but the animals themselves are so crowded that many of them lose their normal shapes and become greatly elongated. When the rocks are covered with water they seem literally alive from the thousands of waving cirripeds or feathery legs.

The shell of the common barnacle is small, white, and variable in shape. Unlike other species the base of the shell is membranous instead of calcareous, a distinguishing feature by means of which the species may be recognized. The opercular valves are also a distinctive feature that serves to identify the species: two are pointed at the tip, two are blunt. The diameter of the shell is less than one-half inch.

A barnacle common from Massachusetts southward to the West Indies is the ivory barnacle, *Balanus eburneus* (Fig. 92). It is found chiefly below low-water mark attached to all kinds of submerged woodwork, whether fixed or floating, and also on the carapace of crabs and on various mollusks. It often occurs in brackish and even in fresh water.

The shell of the ivory barnacle is fairly large, low and broad in form, smooth and white, and unlike the preceding species with a calcareous base. It inclines backward to form a sort of oblique cone; the opening is triangular.

Occurring on stones and shells below low-tide mark, *Balanus crenatus* ranges along the north Atlantic and Pacific coasts. It is a very large species, its shell measuring about one and a half inches in diameter. It is white, cylindrical, rather conical, and rugged with the base calcareous and very thin. The valves of the operculum are acute with diverging points, the latter striated.

An even larger species than the preceding is the purple barnacle, *Balanus tintinnabulum.* The shell is pink or reddish to purple, conical in shape, often ribbed longitudinally, and with a calcareous base. The posterior valves of the operculum are longer than the others and curved forward, resembling the beak of a bird of prey. The species is cosmopolitan but found chiefly in warmer waters. It is often brought to our coasts on the hulls of ships. It is said to serve as food in some countries.

A species found on whales, especially humpback whales, is *Coronula diadema* of the same family as Balanus. The shell is crown-shaped, wider than high, and has a membranous base. A very small and fragile barnacle is *Chthamalus fragilis* (family Chthamalidae). Its shell is conical, rounded or oblong, smooth and light bluish gray or olive brown in color. This species occurs on rocks in the tidal zone but unlike others does not form large clusters but is found singly or in very small groups. Its range is along the Atlantic coast from New England to the West Indies.

A rather odd barnacle is the species *Alcippe lampas.* This animal does not have a shell but has a segmented body that is protected by a mantle, which is a fold of the integument. It has only three pairs of feet on the posterior part of the thorax and a weak stalk that attaches the animal to an object by means of a large, chitinous disc. The male is minute, doesn't have any legs, and is attached to the female. Alcippe bores into dead moon shells that have been appropriated by hermit crabs and also attacks the shells of other barnacles.

It may be somewhat surprising to learn that there are a number of parasitic species of barnacles that are very degenerate. They lack any body segmentation, appendages, or shell, and are sac-shaped with a stalk composed of branched, threadlike projections that extend rootlike into the body of the host. They also lack a digestive tract, their food being absorbed from the tissues of the host through the rootlike extensions of the stalk. These parasitic barnacles are hermaphroditic with complementary males. The best-known genera are Sacculina with about six species that live on crabs and Peltogaster with about seven species that live on hermit crabs.

9 The Mysids

Opossum Shrimps Most of us have probably never heard of the mysids or opossum shrimps, since they are not generally known, and yet they are an important group of animals, since they are valuable as food for fishes and whales, sometimes forming as much as eighty percent of the stomach contents of such fish as the lake trout.

They are shrimplike animals with an elongate body that is usually more or less transparent, a short, thin carapace, and eight pairs of rather feeble, leglike biramose thoracic appendages, none of which are walking legs and none of which (usually) bear claws, the anterior two pairs being slightly modified to form maxillipeds. The eggs are carried in a pouch beneath the thorax and hatch in some species as nauplii. Most of the opossum shrimps live in the sea, only one species being commonly found in the fresh waters of North America, although two are known from brackish and coastal waters and several from caves. They feed, for the most part, on plankton and detritus, which they filter out of the water with their maxillipeds, and in turn serve as food for fishes, as I have already remarked. The mysids number about three hundred species, contained in three families, the family Mysidae (from the Greek meaning a closing of the eyes or lips) having the best-known species. The members of this family have no gills, the maxillipeds are shorter than the following appendages, and there is an auditory sac on each of the six pairs of pleopods or abdominal appendages. There are some twenty-one genera and about ninety species in the family of which the following are representative.

Our sole freshwater species is *Mysis relicta,* which is found in the cold, deep waters of the Great Lakes, the Green Lake of Wisconsin, and the Finger Lakes of New York State. It also occurs in some of the Canadian Lakes and in the lakes of Europe and Asia. It is a slender animal, about five-eighths of an inch long, with a body compressed from side to side. In warm weather it migrates to the deeper parts of the lakes where temperatures may be as low as 4° C., returning to the surface when night begins to fall and remaining there possibly for a few hours if the temperature does not rise above 20°C. Mating takes place during the colder months, and the eggs are carried in a brood pouch or marsupium. Here the eggs hatch and the young are carried within the brood pouch from one to three months in the manner of the opossum. Hence their name of opossum shrimps.

Very common in the eelgrass from the New England coast southward is *Mysis stenolepsis* (family Mysidae). It is about an inch long (Fig. 93) with a more or less cylindrical thorax and a carapace with a short, blunt rostrum and with its lower anterior margin extended to form a sharp tooth. The antennae are very long and the body bends rather sharply between the first and second abdominal segments. The eyes are large and conspicuous and the telson is longer than the sixth segment. This mysid is translucent or white with a star-shaped black spot on each segment with blackish markings also scattered over other parts of the body. The female is somewhat larger than the male. The species is quite abundant in winter on the shores of still muddy bays and sounds and is an important item in the diet of such fishes as the shad and spotted flounder.

Two other species of opossum shrimps that I might mention are *Erythrops erythrophthalma* and *Heteromysis formosa.* The former is readily identified by its very large ruby-red eyes. It occurs from Massachusetts northward. The latter has the first pair of thoracic legs larger than the others and each ending in a claw. The males are colorless, the females rose-colored. This species is found in eelgrass and sometimes in dead mollusk shells along the coast of New England.

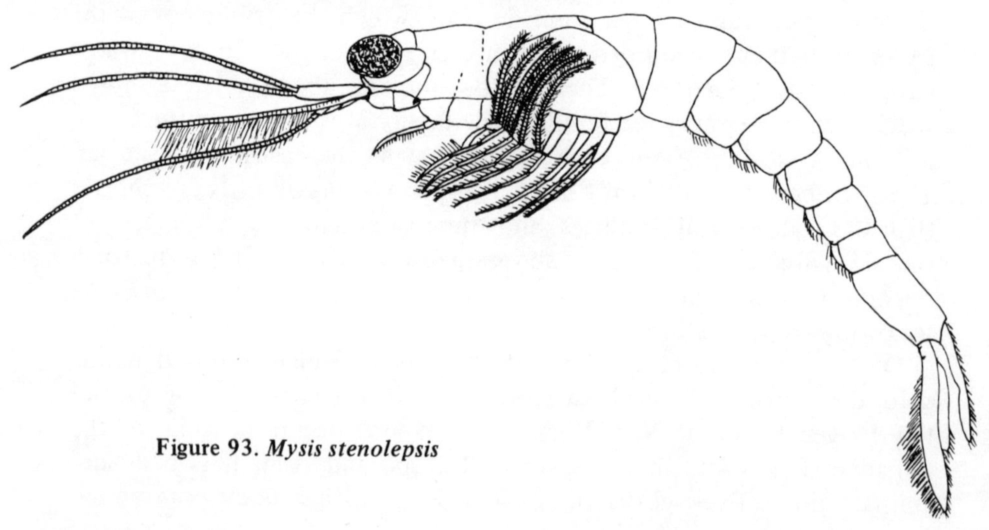

Figure 93. ***Mysis stenolepsis***

10
The Cumaceans

Most of us have never heard of the cumaceans let alone having seen any of them, for they are nearly all marine and live mostly in the sand or mud, though sometimes they swim at the surface when they may be taken with a tow net.

They are small, shrimplike crustaceans, ranging in size from two-eighths to five-eighths of an inch in length. They have a short carapace that does not cover the posterior four or five thoracic segments and a long, slender, more or less cylindrical abdomen. Their first antennae are short, the second are short in the female and long in the male, and their eyes are wanting or when present are set close together and sessile. There are two pairs of maxillipeds and six pairs of periopods or walking legs, some of which are biramous. The pleopods or swimming legs and the uropods are lacking in the female, while the male may have two to five pleopods. A single pair of gills are present on the first pair of maxillipeds. The eggs, which are fairly large, are carried by the female in a brood pouch beneath her thorax, and the young when hatched are like the parents in appearance except that they lack the last pair of thoracic and all the abdominal legs.

Abundant on soft muddy bottoms, in two to two hundred fathoms, from Nova Scotia to New Jersey, *Diastylis quadrispinosa* (family Diastylidae) may be recognized by its pale flesh color and the red-purple spot at the posterior part of the carapace. It is about a quarter of an inch or less in length and may be further distinguished by its large carapace, its posterior marked off in transverse ridges and its anterior extended into a sharp, pointed rostrum. It also has a long, slender abdomen that terminates in long, forked spines that are longer than the telson (Fig. 94).

A species found off the coastal waters of New England, *Cyclaspis varians* (family Bodotriidae) has a large carapace that viewed from the

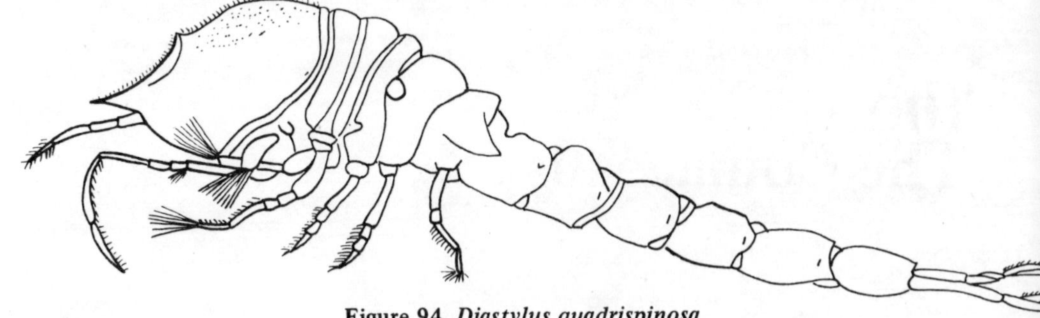

Figure 94. ***Diastylus quadrispinosa***

side seems almost round. The species has a short rostrum and a long and slender abdomen that is much longer than the rest of the body (Fig. 95). The male measures about one-sixth of an inch long, the female one-eighth of an inch. A third species I might mention is *Eudorallopsis deformis* (family Leuconidae). This species occurs from Nova Scotia to Vineyard Sound. It is short and stout and lacks a telson. There are about four hundred species of cumaceans and are doubtless of more interest to the zoologist than to the general reader.

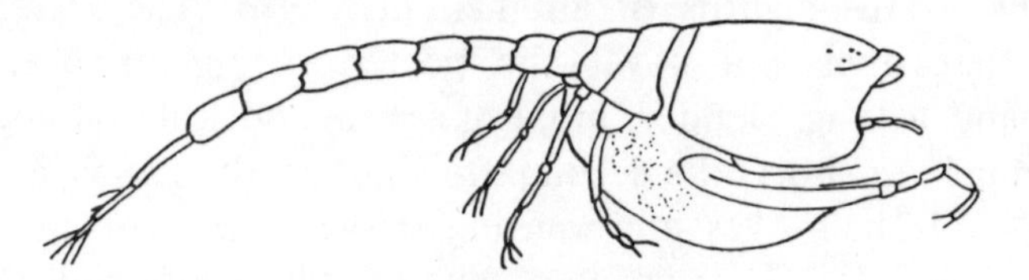

Figure 95. ***Cyclaspis varians,*** **Female**

11
The Isopods

If the cumaceans, which I discussed in the last chapter, appeared to be rather uninteresting animals and partly, I suspect, because few of us ever get to see them, the opposite should be true of the isopods because these animals, or at least many of them, occur in places where we can easily get to view them, as in our gardens for instance or even in our cellars.

The isopods (from the Greek prefix *iso,* meaning equal and from the Latin word *poda,* meaning foot) are a large and widely distributed group of crustaceans, numbering over three thousand species. They vary in size, though they are usually small, and are rather inconspicuous creeping or swimming animals, somewhat retiring in habits and extremely ferocious. Many are marine, either swimming in the open sea, living at great depths, or hiding under stones along the seashore or among the seaweed in shallow water; others inhabit freshwater ponds and streams; and a few are terrestrial, being found under the leaves in the woods or under almost any pile of rubbish among decaying vegetable matter. A few species of isopods are parasitic on fish and decapods, and it is not unusual to find a prawn with what appears to be a very swollen throat but that actually is a parasitic isopod attached to its gills.

The isopods have an elongated, flattened, and more or less arched body, with a thorax of seven segments and an abdomen of six, those of the abdomen often united, sometimes into a single piece. They lack a carapace. The appendages are well developed and when all are present include two pairs of antennae, one pair of mandibles, two pairs of maxillae, one pair of maxillipeds (the latter attached to the first thoracic segment, which is fused with the head), seven pairs of thoracic legs or periopods, which are not equal as the name isopod would imply, but vary in different species, and six pairs of abdominal appendages or pleopods, the anterior pairs of which function as gills, the uropods as feelers. As a rule the thoracic legs are adapted for walking or attachment, the first two pairs often with claws and used for grasping food. The eyes are sessile. The females carry their eggs in a brood pouch on the lower surface of the thorax and the young animals resemble their parents in form and appearance. The isopods are generally scavengers.

A species of isopod found from Greenland to Long Island Sound as well as in Europe and occurring in shallow water on piles and among seaweed is *Tanais cavolini* (family Tanaidae). It is about three-sixteenths of an inch long, with a slender body and an abdomen composed of five segments, the first three being fringed with setae, and three pairs of pleopods, and is brown in color, paler beneath. This species is also found on sponges, barnacles, and oysters. Also of the same family is *Leptochella savignyi* (Fig. 96), which occurs abundantly in eelgrass and

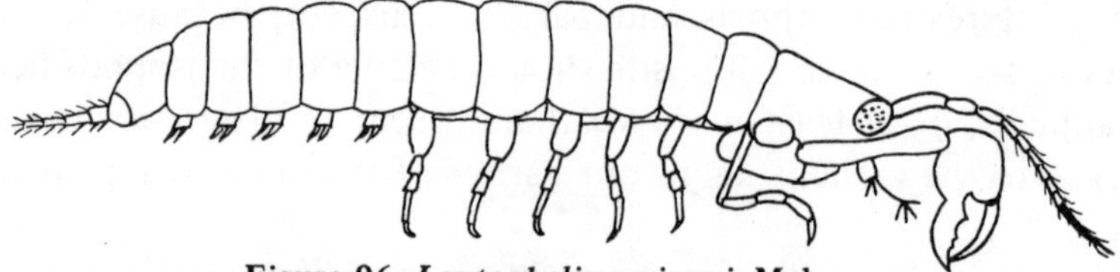

Figure 96. ***Leptochelia savignyi,*** **Male**

seaweed near the sea-surface from Cape Cod to New Jersey. It measures about one-sixteenth of an inch in length, is white in color, and has an abdomen of six segments, the last one triangular, and five pairs of pleopods. The claws of the male are large, those of the female are small.

Rather common between the tide marks or around the low-water mark, where it lives on the sandy or muddy bottoms, *Cyathura carinata* (family Anthuridae) ranges from Greenland to New Jersey and is also found in Europe. It is brownish or yellowish in color with a cylindrical and elongated body and a relatively short abdomen. The first segment of the thorax is longer than the others with a stout pair of legs that end in a fingerlike process with a slender, curved spine. The other six pairs of legs are similar to each other and are used for walking. The telson, together with the uropods, forms a fanlike structure.

The isopod *Cirolana concharum* (family Cirolanidae) is a rather large species, measuring from an inch to one and a half inches in length. It has an almost rectangular head, with the first three pairs of legs adapted for grasping, the remaining four for walking, and with the telson forming with the uropods a swimming fan, both the telson and uropods being fringed with setae. The body is semicylindrical and broad and consists of fourteen segments, the first being the head, the next seven the thorax, and the remaining six the abdomen. In color this isopod is yellowish, with a brown edge on the posterior margins of the segments. It occurs from Nova Scotia to South Carolina, usually in shallow water where it swims about, and feeds on dead fish, as well as upon the blue crab and other animals.

The Gribble An isopod that has a taste for wood and commonly called the gribble is *Limnoria lignorum* (family Limnoriidae). It is a fifth of an inch long and to the naked eye looks like a small, white seed, but its back is covered with minute hairs to which dirt usually clings, thus often hiding its true appearance. The gribble has a flattened body with parallel sides with the segments so well articulated that it can roll itself up into a ball. The antennae are short, as is the head, the eyes are lateral and are set wide apart, and the legs are all uniform in character and are well adapted for walking (Fig. 97).

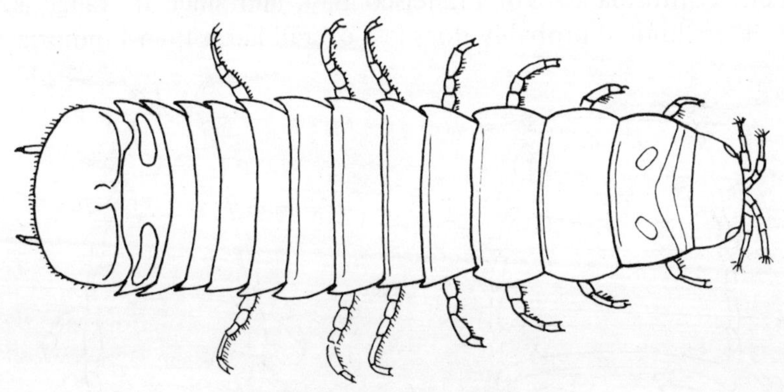

Figure 97. ***Limnoria lignorum***

This isopod is found along both the Atlantic and Pacific coasts of North America and along the coast of Europe. Its natural habitat is in the neighborhood of the low-water mark, where it is usually quite abundant, but it often extends its range to a depth of ten fathoms. It is very destructive in its habits, boring into submerged timbers and woodwork and doing considerable damage to wharfs, piles, and the like. It burrows with its mandibles or jaws, which are chisellike, to a depth of half an inch, complctely honeycombing the surface of the wood, which scales off or quickly decays and is then washed away by the waves. A pest it doubtless is, but it redeems itself in also attacking floating and water-logged timber and similar wreckage and thus cleaning up the sea of debris that could be a serious threat to navigation.

The isopods of the genus Sphaeroma get their name from their habit of rolling themselves into a ball when alarmed. There are some thirty American species, *Sphaeroma quadridentatum* (family Sphaeromidae) being a common species found under stones and among algae between tide marks. The body of these isopods is short, oval, and convex and is so constructed as to admit such a singular change in form. When extended the body is an ellipse and in the case of *quadridentatum* measures a little over a quarter of an inch in length. The head is short, the eyes at its outer margin, the second antennae are longer than the first, the legs are hairy and are adapted to walking, the abdominal appendages are platelike and fringed with hairs, and the abdominal segments are fused into one (Fig. 98). The color of this species, which ranges from Cape Cod to Florida, is variable, some individuals being a uniform slate gray, others marked with a longitudinal patch of color on the back. A somewhat larger species, *Sphaeroma destructor,* varies in color from yellowish to dark brown and has the upper surface covered with granules. It is a boring isopod and is more destructive than the gribble, since its holes are larger. It occurs along the coast of Florida and southern California to San Francisco Bay, and since its range is thus more or less limited probably does less overall harm than Limnoria.

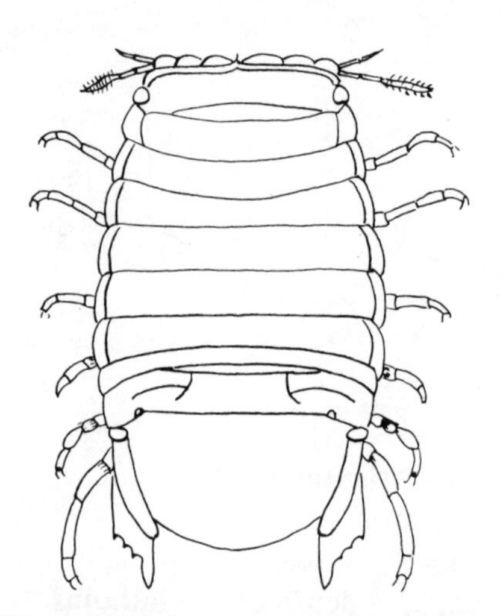

Figure 98. ***Sphaeroma quadridentatum***

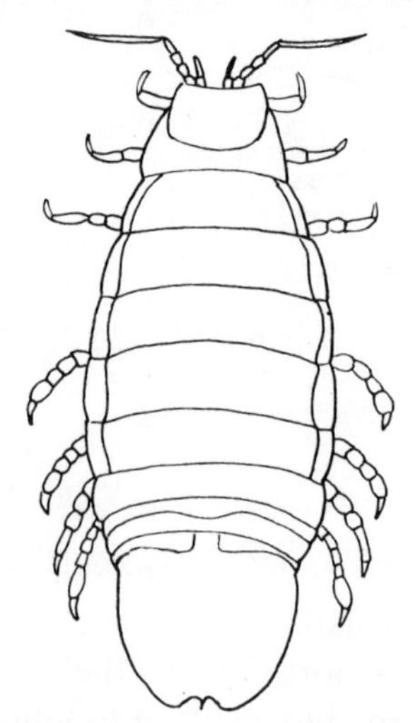

Figure 99. ***Idothea baltica***

A cosmopolitan species, *Idothea baltica* (family Ithotheidae) may be recognized by its abdomen the first three segments of which end in acute teeth and the last three of which are fused together and end in three teeth, the middle one being the longest (Fig. 99). The body of this isopod is more or less broad and flattened, the second antennae are larger than the first, and each has a long flagellum, and the legs are similar to each other. Its color is variable but usually green, though sometimes brown with black spots. On the Atlantic coast of America it ranges from Nova Scotia to North Carolina and may be found on floating seaweed, on the seaweed along rocky shores, and even in the sand. It is about an inch long.

Another member of the same genus, *Idothea metallica,* is slightly smaller and has the end of the abdomen truncated instead of toothed. It is bright blue or green in color, often with a metallic luster when seen in the water, and is found swimming free in open water or in masses of floating seaweed. It ranges from Nova Scotia to North Carolina but is cosmopolitan in distribution. A third species, *Idothea ochotensis,* which is similar to *baltica* occurs along the Pacific coast, from Alaska to San Francisco Bay.

Visit a New England sea beach and you will find the little ridge burrows of the isopod *Chiridotea caeca* (family Idotheidae) in the wet sand below the high-tide mark. This species has a broadly ovate, flattened thorax, with the abdomen nearly half as long as the whole body, and tapering to an acute point at its posterior end. The head has a deep cleft or notch on each side, the eyes situated just back of these notches, the antennae are of nearly equal length, the second usually with a short flagellum, the first three pairs of legs are grasping, the last four walking, and the abdomen has four segments (Fig. 100). This

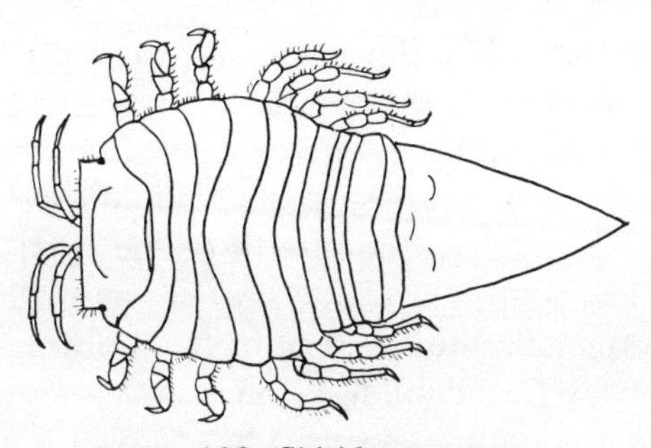

Figure 100. ***Chiridotea caeca***

isopod is about half an inch long and is variable in color, but usually a dark mottled gray. Though commonly found on the sea beaches of New England, where its trails may easily be seen in the wet sand, it ranges from Nova Scotia to Florida. A somewhat similar species, *Chiridotea tuftsi,* but smaller in size and a light reddish brown in color, speckled and marked with darker patches, also occurs on sandy shores but more rarely then the preceding. Its range extends from Nova Scotia to Long Island Sound, from shallow water to twenty-five fathoms.

Another species of the same family is *Edotea triloba,* which may be found under stones and decaying algae in muddy places along the shore. It has an ovate body with conspicuously lobed thoracic segments that are rounded at their outer margins and that give the animal a scalloped appearance. It is a little over a quarter of an inch long and has a uniform muddy color that blends with its environment. Its range is from Maine to New Jersey.

The Salve Bug Called the salve bug because it is used as a salve by fishermen, the species *Aega psora* (family Aegidae) is a parasite on such fishes as the skate, cod, and halibut as well as on other fishes. The body of the salve bug is oval or elliptical in outline, with the upper surface somewhat arched and with very tiny scattered dots. The head is somewhat triangular in shape, the eyes are large and kidney-shaped, the second antennae are longer than the first, the first three pairs of legs are grasping, ending in hooks, and the last abdominal segment is large, triangular, and pointed at the tip. The salve (Fig. 101) bug is distributed along the entire Atlantic coast from Greenland to Florida and is also found in the Gulf of Mexico and around the British Isles.

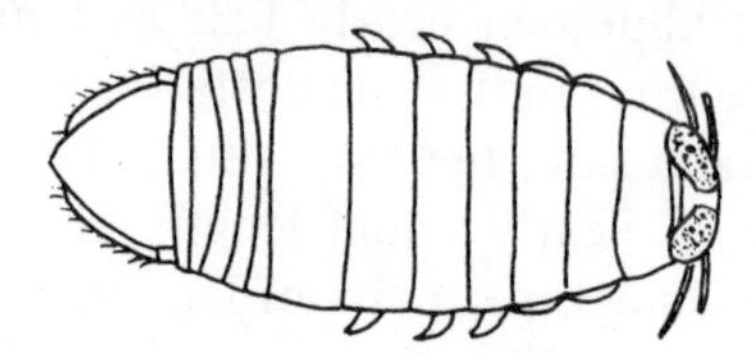

Figure 101. ***Aega psora***

Another parasitic isopod is *Livoneca ovalis* (family Cymothoidae). It has an elliptical, more or less asymmetrical body, a triangular head, large eyes, and all legs ending in curved hooks, which are thus adapted for clinging to fishes. This species occurs from Massachusetts to Florida and in the Gulf of Mexico to the mouth of the Mississippi River and is parasitic on bluefish, scup, sea trout, sawfish, and other fishes, usually attaching itself to the gills and the roof of the mouth.

A common species found under stones and seaweeds between tide marks, *Jaera marina* (family Janiridae) has an oval, somewhat flattened body with the sides of the head expanded as lamellae under the eyes, and very small uropods and first antennae. It is about one-fifth of an inch long and though very variable in color is usually a mottled gray. Its range extends from Labrador to southern New England, and it is also found along the coast of Europe.

Although the family Asellidae includes some marine species, the members of the family are essentially freshwater animals. They have a flattened body with seven free thoracic segments, the segments of the abdomen forming a single, shieldlike plate, and their abdominal appendages are modified as respiratory organs.

The most common freshwater isopod is *Asellus communis* (Fig. 102). It is about half an inch long, with the first antennae short, the second long, and the first pair of legs prehensile, the other ambulatory. Asellus can often be found in pools that seem altogether unpropitious for animal life and where, as a matter of fact, there is hardly anything else living. In such places it crawls over the muddy bottoms feeding on dead

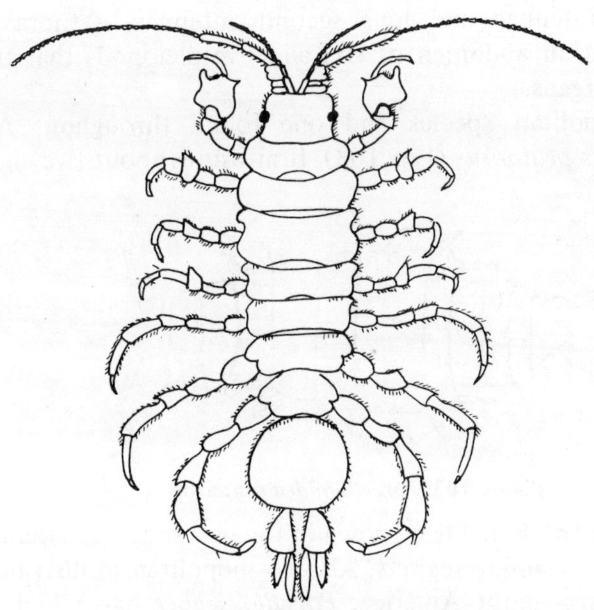

Figure 102. *Asellus communis*

leaves and refuse. It is more or less gray or grayish brown in color and when seen from above looks like a miniature armadillo. From very early spring and continuing through the summer the females have a new brood of young every five or six weeks; indeed, they always seem to be carrying a brood pouch full either of eggs or young.

During the fall the males and females seem about equal in number, but from midwinter on the males become more numerous. In winter when conditions are unusually favorable Asellus will gather in very large numbers, thousands having been observed in slow streams beneath the ice, some swimming with their heads pointed upstream, others resting in groups on the bottom, and many being swept downstream by the current, rolling along the bottom in balls of from a few to many individuals. In early spring Asellus may be seen in incredible numbers.

Asellus is seldom seen in open water, preferring to hide in vegetation or under logs and rocks of ponds and pools, often in temporary pools that dry up in summer, and even in acid water where insects are scarce. It can withstand stagnation. A related species, *Asellus stygia,* lives in caves and deep wells.

Sow Bugs The isopods we find in our gardens, as I have already mentioned, and in the woods beneath logs and in damp dark places belong to the family Porcellionidae and are generally known as sow bugs, though they are neither bugs nor insects. They have a depressed, oval body, a small head that is sunk into the first thoracic segment, small first antennae and long second antennae, a thorax of seven segments and an abdomen of six, and five pleiopods that function as respiratory organs.

A cosmopolitan species and one found throughout America is *Porcellionides pruinosus* (Fig. 103). It measures about five-eighths of an

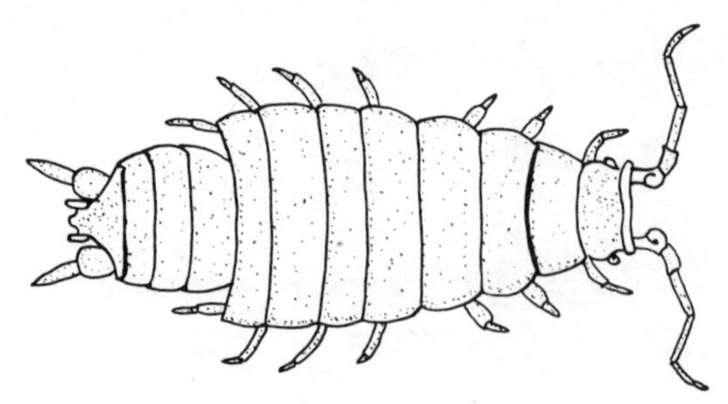

Figure 103. *Porcellionidea pruinosus*

inch in length and is reddish brown in the posterior and lateral regions, lighter in the remaining regions. Also cosmopolitan in distribution and also found throughout America, *Pocellio scaber* has a body covered with minute tubercles. It is about one-half inch in length, is uniformly black in color, and occurs under bark, logs, and similar objects. Found in crevices of rocks and in damp places along the Atlantic slope and in Europe, *Porcellio spinicornis* is yellowish gray in color variegated with dark brown patches. *Trachelipus rathkei* has black blotches on a yellowish-brown ground color, with two lateral and usually a median light stripe. It is common under boards, stones, and similar objects throughout the eastern and central states as well as in Europe. A brown or dark gray species that is spotted with white is *Cylisticus convexus.* It has a somewhat elongated, smooth, very convex body, long uropods, and a head with lateral lobes and is able to roll itself into a ball. It has the same habitat and distribution as *Trachelipus rathkei.*

Wood Lice Turn over a fallen log in the woods and you will likely see small, slate-gray animals running about in bewilderment. These crustaceans are popularly called wood lice and are members of the family Oniscidae. They are oval in form and somewhat flattened and with the same structural characteristic as the members of the Porcellionidae. They are quite common under logs and in other dark, damp places. One of the more abundant species is *Oniscus asellus* (Fig. 104). It is about five-eighths of an inch long, deep slate in color and spotted with white. It is found throughout the eastern and central states as well as in Europe. *Philoscia muscorum* is about half as long and is usually dark brown in color with two darker median stripes. It is found under stones and boards above high tide along the seashore from Cape Cod to New

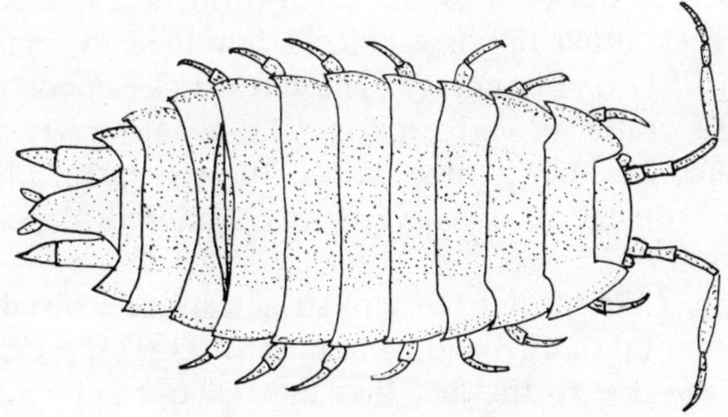

Figure 104. *Oniscus asellus*

Jersey. *Trichoniscus demivirge* has an elongated, elliptical body with the abdomen abruptly narrower than the thorax. It occurs in eastern North America where it may be found beneath moss or dead leaves in the woods, and is one of four species of the family Trichoniscidae.

Pill Bugs Perhaps you may have heard of the pill bugs, which, like the sow bugs, are neither bugs nor insects, but isopods belonging to the family Armadillidiidae. They have a convex body, minute first antennae, and short uropods, and are able to roll themselves into a ball (although other isopods can do so too), hence their name. *Armadillidium vulgare* is a cosmopolitan species and is found all over America. It is about five-eighths of an inch long and is black or dark gray with rows of indistinct spots (Fig. 105). Also cosmopolitan in distribution is *Ligyda exotica* (family Lygydidae). It has an elliptical body and is rather common among rocks and on piles and docks along the shore.

From early spring to late autumn the wood lice may be observed creeping about their interlacing runways beneath any board or log that has lain for some time on damp ground. They are abundant on the forest floor and in many places hardly a stick can be overturned or the dead leaves be scattered without disturbing them. They live on organic material in the soil and on decaying plant matter and are able to survive under the poorest of living conditions. In their natural habitats these isopods cluster together when conditions become dry in order to conserve their own moisture and to reduce the effect of temperature

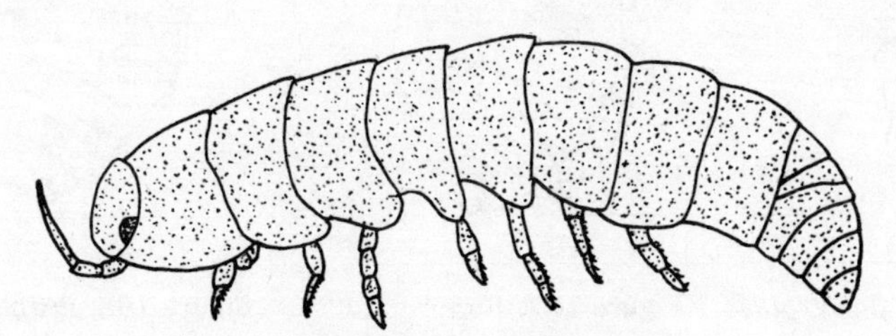

Figure 105. *Armadillidium vulgare*

changes, much as honey bees and earthworms do. It seems that when they touch each other or some object, they tend to remain quiet and thus expend little bodily energy. The terrestrial isopods produce their first brood of young in early spring. The females carry them in their brood pouches for about three weeks, the young ranging in number from about a dozen to as many as two hundred, depending on the species.

It seems most likely that the terrestrial isopods evolved from marine forms that went up on the land from the sea. Their success in remaining on the land was due to the fact that their pleopods could be modified for breathing atmospheric oxygen and that other adaptations could be developed for water regulation involving the integument, excretory organs, and the like. But at best the terrestrial isopods are imperfectly adapted to land since they cannot for long survive in dry air. In other words they require an ecological niche in which the humidity is fairly high or where there is available water.

The isopods belonging to the family Bopyridae are said to be the most degenerate members of this very large group of crustaceans, all being parasites on the decapods. The males and females are dissimilar, the females being asymmetrical and broad, sometimes greatly deformed, in some species being merely a simple sac with eggs; the males are smaller and more slender, and symmetrical. The antennae are rudimentary and the legs are prehensile, though in some species they are rudimentary on one side.

The species *Bopyroides hippolytus* (Figs. 106, 107) is circumpolar in distribution, its range extending to Boston and Puget Sound. It is parasitic on the gills of various prawns, its presence often showing as a swelling under the carapace. *Probopyrus pandalicola* (Fig. 108) is widely distributed along the northern border of North America on both the Atlantic and Pacific coasts and also around the North Atlantic to

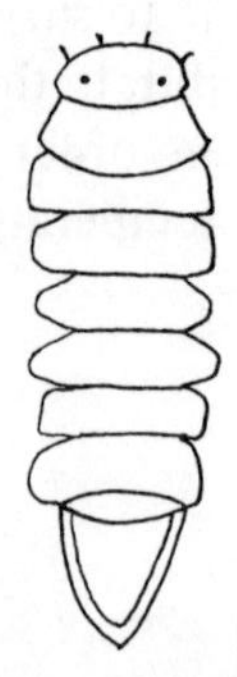

Figure 106. ***Bopyroides hippolytes,*** **Male**

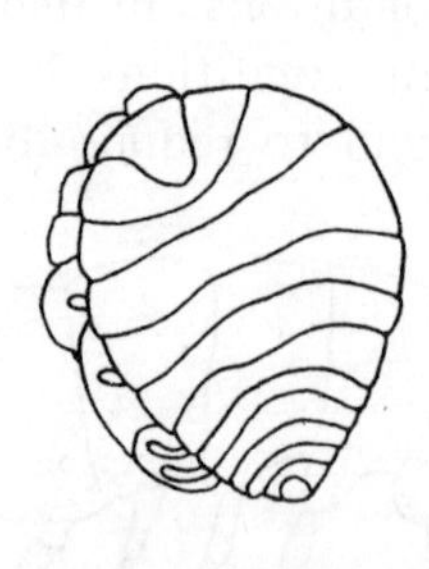

Figure 107. ***Bopyroides hippolytes,*** **Female**

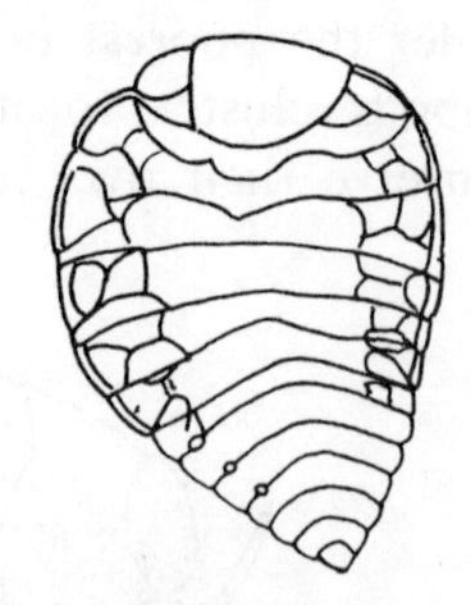

Figure 108. ***Probopyrus pandalicola***

the coast of Europe. The female, white in color with black markings, lies against the body of the host; the male usually is found clinging to the female. It is a parasite of the common prawn as well as of other species of prawns. Other parasitic isopods are *Phryxus abdominalis,* a parasite of the hermit crab, and *Leidyi distorta,* a parasite of the fiddler crab.

12
The Stomatopods

If you are familiar with the praying mantis you should have no trouble in recognizing the mantis shrimp, for the two resemble each other in form, as well as in their savagery and in their manner of eating.

The mantis shrimp belongs to a group of crustaceans known as the stomatopods (Stomatopoda, *stomato,* a combining form from the Greek meaning mouth, and the Latin *poda,* foot). They are rather large animals, often reaching a foot in length, with a small, flat carapace that does not cover the posterior thoracic segments and a broad elongated abdomen, its appendages bearing gills. The telson is rather expanded, forming the widest part of the body, and is usually armed with three pairs of large spines. There are five pairs of thoracic appendages that are modified as maxillipeds, the second pair being much larger and heavier than the others, being specialized to form raptorial organs with which the stamatopods seize their prey. The pleopods are biramous and powerful swimming organs. The heart is long and tubular and the liver, testes, and ovaries extend the length of the thorax and abdomen. There are about two hundred species contained in one family, Squillidae, and all are marine, living in holes in shallow water.

The mantis shrimp *(Squilla empusa)* mentioned above measures from eight to ten inches in length and about two inches in width (Fig. 109). The carapace has a longitudinal median ridge with two other ridges on either side, and there are four pairs of ridges on the first five abdominal segments. The terminal joint of the second maxillipeds has six long teeth that fit into sockets in a groove on the second joint. By means of this singular organ the mantis shrimp can hold onto its prey securely and can inflict a severe wound in the human hand if handled carelessly. In color it is pale yellowish green, each segment being bordered with darker green and edged with yellow. The mantis shrimp occurs from Cape Cod to Florida and lives in burrows in the mud between tide

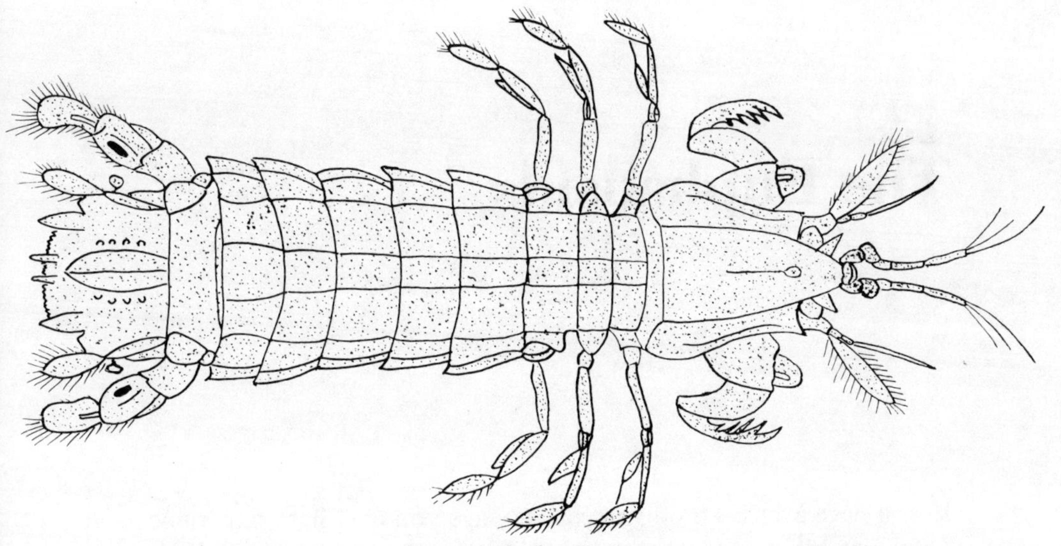

Figure 109. *Squilla empusa*

marks and below the low-water mark, each burrow usually with several openings a few feet apart.

There are several other stomatopods that bear mention such as *Pseudosquilla ciliata,* which is mottled dark green and brown and which occurs in shallow water off the southern coast of Florida; *Lysiosquilla scabricauda,* found in the Gulf of Mexico from Florida to Texas; and *Gonodactylus aerstedi,* which is generally green and brown but variable in color and is found from North Carolina to the Gulf of Mexico. The last species is very abundant in the West Indies. These three species have the general characteristics of the family but differ in minor structural details.

13
The Euphausids

If you have ever been on an ocean voyage you may have experienced an unforgettable sight, that of having seen a luminescent sea, such luminescence being due to the euphausids, shrimplike animals that have small spherical organs capable of emitting considerable luminous power, their light being directed and controlled by a complicated apparatus. Place a dozen or so of these animals in a glass jar and you will have enough bluish white light to read by.

The euphausids (their name means shining light) have two-branched appendages, with no special maxillipeds, all the thoracic appendages or periopods being walking legs. The abdominal appendages or pleopods function as swimming legs. The gills are attached to the periopods. They have a transparent body that is highly luminescent at night, the phosphorescence being highly developed on the outer margins of the eye-stalks, on the bases of the second and seventh periopods, and on the ventral median line of the first four abdominal segments. They are mainly colorless but some have spots or washes of pink or red.

Most euphausids live between the surface and three thousand feet but some range to depths of sixty-five hundred feet. They prefer the cold seas, but there are some that are found in deeper, warmer water. At times they congregate in northern seas and in the Antarctic Ocean in such numbers that the water turns red, when it is often referred to as "tomato soup." The euphausids that occur in the warmer waters are carnivorous in their food habits, but most live on floating plants that they filter out with the feathery hairs on their front legs after first having set up a backward flowing current by rhythmically moving their abdominal legs. The egg pouch is under the rear of the thorax and the young are hatched as nauplii. The euphausids are consumed in vast numbers by whales and other swimming creatures. Whalermen refer to the euphausids as "krill" and have long known that when the krill put in an appearance whales will soon be sighted.

There is but a single family of these shrimplike crustaceans, Euphausiidae. A species that occurs from Massachusetts Bay and off the Maine coast to the Bay of Fundy is *Nyctiphanes norvegica* (Fig. 110). It is translucent and has black eyes. Both the body and appendages are tinged and marked extensively with red. The name Nyctiphanes is derived from two Greek words meaning "the night shining one" and refers to its brilliant luminescence. *Rhoda inermis* is a similar species but of paler coloration and greater transparency. It is a northern species occurring on both sides of the Atlantic. An American species, *Euphausia pacifica,* is common off the coast of southern California.

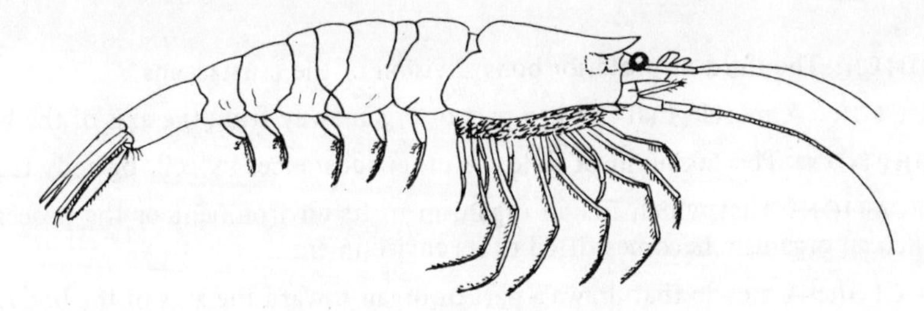

Figure 110. ***Nyctiphanes norvegica***

Glossary

ABDOMEN: The third or posterior body division of the crustaceans.

ABDUCTOR: A muscle that draws a part or organ away from the axis of the body.

ABSORPTION: The taking in of fluids or other substances by cells or body tissues.

ADAPTATION: That which fits an organism to its environment or the process by which an organism becomes fitted to its environment.

ADDUCTOR: A muscle that draws a part or organ toward the axis of the body.

ALGAE: Very simple green plants.

ALIMENTARY TRACT: The digestive canal or organ that ingests, digests, and absorbs the food.

ANALAGOUS: Having a similar function, that is, parts or organs with a similar function.

ANTENNA: A segmented sensory appendage of the head; a movable sense organ on the head.

ANTENNAL SCALE: The exopodite of the second antenna in certain crustaceans.

ANTENNAL SINUS: An indentation in the shell near the antennae of certain ostracods.

ANTENNULE: A many-segmented sensory organ of the head located near the antenna and often called the first antenna.

ANTERIOR: Pertaining to the front or head end of an animal.

ANUS: The posterior opening of the digestive canal.

APERTURE: An orifice or opening.

APPENDAGE: A portion of the body that projects and has a free end, such as a limb for instance.

AQUATIC: Pertaining to water; living in water.

ARTERY: A blood vessel that carries blood away from the heart.

ARTHROPOD: An invertebrate animal with segmented or jointed appendages.

ARTICULATE: Composed of a series of homologous segments.

ARTICULATION: A joint as between two segments or structures.

ASEXUAL REPRODUCTION: Reproduction not involving sex cells.

ASYMMETRY: The condition in which opposite sides of an animal are not alike.

AUDITORY: Pertaining to the sense or organ of hearing.

AURICLE: A chamber of the heart that receives blood from the veins.

AUTOTOMY: Self-mutilation. The automatic "voluntary" breaking off of a part of the body of an animal.

BEHAVIOR: The reactions of an entire organism to its environment.

BENTHOS: The organisms living on or in the bottom of a body of water, either salt or fresh, from the edge of the water to the greatest depths.

BILATERAL SYMMETRY: Having both sides alike, that is, the right and left sides of the body. The arrangement of the parts of an organism in such a way that the right and left halves of the body are the mirror images of each other.

BIRAMOUS APPENDAGE: A structure with two branches.

BIVALVE: A shell composed of two distinct and equivalent parts or valves.

BRANCHIA: Gills.

BRANCHIAL: Relating to the gills.

BROOD SAC: A chamber in certain crustaceans in which the eggs develop.

CALCAREOUS: Formed of carbonate of lime or calcium carbonate.

CARAPACE: The shell covering a part of all of the cephalothorax.

CARINA: The median dorsal shell in barnacles.

CARNIVOROUS: Eating other living animals.

CAUDAL: Pertaining to the tail or posterior part of an animal.

CEPHALOTHORAX: A body division formed from the fusion of the head and thorax.

CERCOPODS: The paired bristles in some crustaceans extending from the telson.

CHARACTERISTIC: A distinguishing structure or function.

CHELATE: Having forcepslike pincers.

CHELIPED: The large grasping claw in many curstaceans.

CHITIN: A complex hard and very resistant organic substance present in the cuticula of many arthropods.

CILIA: The numerous vibratory projections on both the outer and inner surfaces of many animals. The microscopic, hairlike protoplasmic processes projecting from the free surface of certain cells and capable of vibration.

CIRRI: Small, slender projections or appendages appearing almost like tentacles except for their position.

CLASS: A main subdivision of a phylum.

CLONE: The offspring produced by asexual reproduction of a single animal.

COLONY: A group of individuals, unicellular or multicellular, of the same species that have developed from a common parent and remain organically attached or held together.

COMMENSALISM: An association of individuals of two different species in which at least one is benefited and the other is neither benefited nor harmed.

COMMUNITY: A more or less complex group of plants or animals that occupy a particular area.

COMPLEMENTARY MALE: A minute accessory male animal in certain barnacles.

COMPOUND EYE: An eye made up of a number of separate elements or ommatidia in arthropods.

COPULATION: Sexual union of two individuals involving the transfer of sperms from the male to the female body.

CORNEA: The outer, transparent layer of the eye.

COXA: The proximal segment of an arthropod's leg by which it articulates with the body.

CUTICLE: The thin, noncellular outermost covering of an organism.

CUTICULA: The outer layer of the integument of most invertebrates.

DEVELOPMENT: The series of changes in the early life of an animal by which it passes from the condition of a fertilized egg to that of the adult.

DIGESTION: The conversion of complex unabsorbable food materials into soluble forms that may be absorbed.

DISTAL: The position or a part of an organ away from the point of attachment; opposed to proximal.

DIURNAL: Pertaining to the time of daylight; pertaining to day.

DORSAL: On or toward the back; pertaining to the back.

ECDYSIS: Molting; the shedding of the outer cuticular covering of an arthropod.

EGG: The nonmotile gamete developed by the female; a female sex cell.

EMBRYO: A young animal that is passing through its developmental stages, usually within the egg membranes or within the maternal uterus.

ENDOPODITE: The innermost of two terminal branches of the typical crustacean leg.

ENDOSKELETON: A supporting structure on the inside of an animal, whether it be cartilaginous, bony, or some other material.

ENVIRONMENT: The place where a species of animal is found in nature and the conditions that are present. The total of physical, chemical, and biological conditions that surround an organism.

EPHIPPIUM: The shell in which the winter eggs of the Cladocera are often contained.

EPIDERMIS: The outer cellular layer or layers covering the external surface of a metazoan; it secretes the cuticle in some animals.

EPIPODITE: A long, slender structure fastened to the protopodite of a walking leg of a crustacean.

EXCRETION: The discharge of metabolic wastes; also the substances discharged.

EXOPODITE: The outermost of two terminal branches of the typical crustacean leg.

EXOSKELETON: A supporting structure on the outside of the body of an animal.

FACET: The external surface of an individual ommatidium.

FAMILY: The principal subdivision of an order.

FECES: The indigestible, unabsorbed residue of digestion.

FERTILIZATION: The union of a mature sperm and a mature ovum to form a zygote.

FLAGELLUM: The terminal portion of the antenna in crustaceans.

FOSSIL: The remains or other indications of prehistoric forms of life.

FREE-LIVING: Not parasitic or attached.

FRONTAL APPENDAGE: A pair of extra appendages between the second antennae in certain Branchiopoda.

FUNCTION: The action of any part of a plant or animal.

FURCA: A pair of projections at the hinder end of the body of copepods.

GAMETE: A mature reproductive or germ cell; a sperm cell or egg cell.

GANGLION: A group of nerve cells.

GASTRIC: Pertaining to the stomach, as the gastric glands.

GENITAL: Pertaining to the reproductive organs of either sex.

GENUS: The taxonomic subdivision of a family. A genus is usually composed of several species or a number of species. The genus name is Latinized, capitalized, and when printed, italicized.

GERM CELL: Gametes or cells that give rise to gametes.

GILL: An organ for breathing the air contained in water; a type of respiratory organ for aquatic organisms.

GIZZARD: A muscular part of the digestive tract used for grinding ingested food.

GLAND: One or many associated cells that secrete or excrete one or more special substances.

GNATHOPOD: A grasping claw in amphipods.

GONAD: A reproductive organ, either ovary or testis, in which reproductive cells (ova and sperms) are produced.

GREGARIOUS: Living in company, as in flocks or herds.

GULLET: Synonym for esophagus.

HABITAT: The environment in which an animal lives.

HEAD: The anterior body division of an animal.

HEART: A muscular, tubular, or saccular organ that propels the blood through the arteries.

HERBIVOROUS: Feeding chiefly on plants.

HERMAPHRODITIC: Having the two sexes united in one animal; an individual having both the male and female reproductive organs.

HIBERNATION: The passing of winter in a dormant or inactive state.

HOST: The animal that harbors a parasite.

INGEST: To take any substance from the outside, especially food, into the digestive tract.

INTEGUMENT: The outer covering of an animal.

INTERMEDIATE HOST: The animal that harbors the larval form only of a parasite.

INTESTINE: The division of the digestive tract in which absorption takes place.

KIDNEY: An excretory organ, usually the chief organ for the excretion of liquid nitrogenous wastes in vertebrates but also often loosely applied to analogous organs in certain other animals.

LABIUM: The under lip of many arthropods.

LABRUM: The upper lip of many arthropods.

LARVA: A young animal that has left the egg and is leading a free life but that has not yet completed its development. An immature, free-living stage in the life cycle of various animals that reach the adult form by undergoing metamorphosis.

LATERAL: A position to the right or left of the median line. The side of the body.

LUMINESCENCE: The production of light as a result of chemical reaction in cells.

MANDIBLE: The anterior pair of mouthparts in arthropods. A jaw; either jaw of an arthropod.

MARINE: Of or pertaining to the sea, ocean, or other bodies of salt water.

MAXILLAE: The paired mouthparts immediately behind the mandibles in arthropods.

MAXILLIPEDS: The anterior, or one of the first three pairs of, thoracic appendages in the Crustacea.

MEDIAN: Refers to the midline or near the middle of the body.

METABOLISM: The sum total of the reactions, mainly chemical, that occur within the protoplasm of an organism.

METAMORPHOSIS: A marked structural change or transformation during development, as from the larva to the adult.

MIGRATION: Movement of a part of the population of a species from one region to another.

MOLT: To shed the cuticula or the outer portion of it; to cast off the exoskeleton.

MOUTHPARTS: The masticatory appendages on the head of arthropods.

NAUPLIUS: A young larval form of many crustaceans.

NICHE: The sum total of environmental factors into which a species fits or that is required by a species.

NUTRITION: The sum of the processes concerned in the growth, maintenance, and repair of the living body as a whole or of its constituent parts.

OCELLUS: A simple type of eye in many invertebrates, especially in arthropods.

OESOPHAGUS: The gullet, the division of the digestive canal leading from the pharnyx to the stomach.

OMMATIDIUM: A single element of the compound eye of an arthropod; one of the elongated, rodlike units of the compound eye.

OMNIVOROUS: Eating all kinds of food, both plant and animal.

OPERCULUM: A plate closing an opening or covering some other structure.

ORGAN: Any part of an animal performing a definite function; a group of cells or tissues that are associated in the body to perform one or more functions.

ORGANIC COMPOUND: A molecule containing the element carbon.

ORGANISM: Any living individual, either plant or animal.

OVARY: The female sexual gland or gonad in which the eggs (ova) multiply and develop.

OVIDUCT: A tube that conveys the eggs from the uterus or to the exterior.

OVUM: The female sex cell; the egg.

PARASITE: An animal that lives in or on another and feeds upon its nutritive fluids. An organism that lives during the whole or a part of its life upon or within another organism and from which it derives nourishment.

PARTHENOGENESIS: Reproduction by means of unfertilized eggs. The production of offspring from unfertilized eggs.

PEDUNCLE: The stalk by which a sessile animal or a sessile organ is attached.

PERIOPODS: The thoracic appendages posterior to the maxillipeds in crustaceans.

PHARNYX: The division of the alimentary canal immediately back of the mouth. The anterior part of the alimentary canal between the mouth and the oesophagus.

PHYLUM: Any one of the main taxonomic divisions into which the animal kingdom is divided.

PIGMENT: Coloring matter.

PLANKTON: A collective term referring to all small forms of life in the surface waters of the sea or body of fresh water.

PLEOPODS: The abdominal appendages in crustaceans.

POSTERIOR: At or toward the hinder end of an animal. The tail or toward the hind or rear end.

PREDACEOUS: Capturing living animals for food.

PROTEIN: An organic compound always containing oxygen, hydrogen, carbon, and nitrogen and often other elements.

PROTOPLASM: The living substance of which all organisms are composed; it is a complex physicochemical colloidal solution and constitutes the physical basis of life.

PROTOPODITE: The basal segment of a crustacean's leg.

PROTOZOA: The phylum of unicellular animals, though sometimes multicellular when there is no specialization of somatic cells.

PROXIMAL: The position of a part of an organ toward the point of attachment—opposed to distal.

RECTUM: The posterior division of the digestive tract.

REGENERATION: Replacement by growth of a part of the body that has been lost.

RETRACTOR MUSCLE: A muscle that draws an organ toward its point of attachment.

REPRODUCTION: The production by an organism of others of its kind.

RESPIRATION: The actual use of oxygen by the cell.

RETINA: The light-sensitive area or layer of an eye.

ROSTRUM: A projection of the carapace in crustaceans.

SALIVARY GLANDS: Digestive glands at the anterior end of the digestive tract.

SECONDARY FLAGELLUM: A small branch of the tentacle in certain crustaceans.

SECRETION: The production of a substance by the protoplasm that is of use to the organism; also the substance produced.

SEGMENT: One of a number of serial divisions of an animal's body or of an organ.

SESSILE: Fixed to one place; not free or moving.

SETA: A bristle.

SELF-FERTILIZATION: The fertilization of an egg by the sperm of the same individual.

SEMINAL RECEPTACLE: A saclike organ that receives and stores sperms after their release.

SEXUAL DIMORPHISM: Phenomenon of the two sexes of a given species differing in secondary sexual characters.

SEXUAL REPRODUCTION: Reproduction involving sex cells, that is, sperms and eggs.

SOMITE: One of the serial segments of a segmented animal.

SPECIES: A group of plants or animals that have certain permanent characteristics in common and that are reproductively isolated from other such groups.

SPERM: A mature male reproductive cell or gamete.

SPERMATOPHORE: A capsule or mass of spermatozoa.

SPERMATOZOON: The male sexual cell.

STATOCYST: Organ of equilibrium in animals such as the crayfish.

STATOLITH: A solid body within a statocyst.

STOMACH: A division of the digestive tract in which digestion takes place.

SUBCHELATE: A pinching claw formed by the bending of the terminal segment over the next one.

SUMMER EGGS: Thin-celled eggs produced by parthenogenetic females of certain crustaceans, usually in the summer.

SWIMMERET: The abdominal appendage of a crustacean; a pleopod; usually functions as a swimming organ.

SYMMETRY: The state of being symmetrical; an organ can be said to possess symmetry if it can be divided by a line or plane into two parts that are essentially similar.

SYSTEM: A group of organs concerned with the same general function, as circulation or digestion.

TAXONOMY: The science that deals with the classification of organisms.

TELSON: The terminal segment of a crustacean; a terminal extension of the last abdominal segment of a crustacean.

TERMINAL: Toward or at the posterior or distal end.

TERRESTRIAL: Living on the ground or land.

TESTIS: The male sexual gland in which sperms are formed.

THORACIC: Pertaining to the thorax.

THORAX: The body division just posterior to the head.

TISSUE: A group of cells of similar structure that perform a specialized function.

URETER: A tube forming the outlet of a kidney. The tube that carries urine away from the kidney to the urinary bladder or the cloaca.

URINE: The liquid waste excreted by the kidneys.

UROPOD: The sixth swimmeret of certain crustaceans that forms the swimming tail.

VARIATION: Difference in structure or function shown by individuals of the same species.

VARIETY: In taxonomy a division of the species; a group of individuals within a single interbreeding population that differs in some minor respect from the rest of the species.

VAS DEFERENS: A duct leading from the testis toward the external opening; the sperm duct. A duct that carries sperms away from the testis.

VEIN: The vessel that brings blood toward the heart.

VENTRAL: On or toward the underside of an animal; away from the back. Opposite of dorsal.

VESTIGIAL: A degenerate structure that was better developed or functional at one time.

WINTER EGGS: Thick-shelled eggs produced by parthenogenetic females, usually in the fall.

ZOOLOGY: The science of animal life.

ZYGOTE: A fertiliaed egg; the cell that results from the fertilization of the egg cell by the sperm cell.

Selected Bibliography

Arnold, A. E. *The Sea-beach at Ebb Tide.* New York: Dover Publications.

Klots, E. B. *The New Field Book of Freshwater Life.* New York: G. P. Putnam's Sons, 1966.

Miner, R. W. *Field Book of Seashore Life.* New York: G. P. Putnam's Sons, 1950.

Morgan, A. H. *Field Book of Animals in Winter.* New York: G. P. Putnam's Sons, 1939.

Morgan, A. H. *Field Book of Ponds and Streams.* New York: G. P. Putnam's Sons, 1930.

Needham, J. G., and Needham, P. R. *A Guide to the Study of Freshwater Biology.* San Francisco: Holden-Day, 1962.

Needham, J. G. and Lloyd, J. T. *The Life of Inland Waters.* Ithaca: Comstock Publishing Company, Inc. 1937.

Pennak, R. W. *Freshwater Invertebrates of the United States.* New York: Ronald Press, 1953.

Pratt, H. S. *A Manual of Common Invertebrate Animals.* Philadelphia: Blakiston, 1935.

Ward, H. B. and Whipple, G. C. *Freshwater Biology.* New York: Wiley, 1918.

Index